**GEMS-NET**
DR. B.YOUNG · 1996

# *Multicultural Mathematics Materials*

Marina C. Krause

California State University
Long Beach, California

With drawings by the author

D1209416

National Council of Teachers of Mathematics

Library of Congress Cataloging in Publication Data:

Krause, Marina C.
    Multicultural mathematics materials.

    Includes bibliographical references.
    1. Mathematics—Study and teaching.   2. Mathematical
recreations.   I. National Council of Teachers of
Mathematics.   II. Title.
QA16.K7 1983      372.7      83-10613
ISBN 0-87353-206-6

Printed in the United States of America

# Contents

# Preface

The activities and games collected here have their roots in different parts of the world. Many have lasted through centuries. Such activities can bring to the present day mathematics curriculum the vitality of ethnic and cultural diversity. They can further enhance the background of the ethnically "different" child and expose children to the ethnic heritage of others.

The materials are classified by geographic region: Africa, Asia, Oceania, Europe, the Middle East, South America, Middle America, and North America. Certain games, however, are difficult to classify with certainty; evidence of a particular game is sometimes found in more than one part of the world, and time may conceal its actual origin.

The North American region includes a special section devoted to the Hopi Indians of northeastern Arizona. These Hopi materials are the result of a careful search of the literature, firsthand study of artifacts in the Hopi collections of the Southwest Museum in Los Angeles and the Heard Museum in Phoenix, and numerous visits to the Hopi mesas. In my studies I found that mathematical elements have been woven into the Hopi way of life from its prehistoric beginnings. Those identified here can be incorporated in the curriculum at the elementary school level.

Although grade levels are suggested, many of the activities cross grade levels. Depending on the students, some might be used at levels above or below those designated. The game boards and activities are designed for immediate classroom use. Only simple, readily available materials are required.

I first compiled and developed these materials for multicultural mathematics courses for teachers. Those teachers, in turn, tried the activities in their classrooms. Therefore the activities have withstood the ultimate test: classroom use with children. My hope is that you and your students will find them enjoyable not only from a mathematical but from a cultural standpoint.

# EGYPTIAN MATCH

## Background

The ancient Egyptian civilization spanned approximately three thousand years. The Pyramids, the oldest buildings in the world today, have already stood nearly five millennia and will be standing millennia from now. The early Egyptian rulers, or pharaohs, were very powerful; the Pyramids were built as tombs to protect their bodies.

Many theories exist about why the pyramid was selected as the shape for these tombs. Among them is the practical theory that the pyramid was the easiest way to build a large building. Another theory likened the pyramid to a ramp representing the sun's rays: the dead rulers could climb up this ramp to heaven. These great structures serve as silent testament to the mathematical skills of the Egyptians, who did not have metal measuring tapes—they used measuring cords made of palm or flax fiber.

The Egyptian numeration system was additive—the values of the numerals were added together. They used picture symbols, such as ropes, flowers, and fingers. The symbol arrangement did not matter, and the ropes and flowers could be turned either way. When multiple use of a symbol was involved, the symbols were arranged in sets of three.

## Materials

A design or grid that can be partitioned into two matching sides. (This design or grid can be drawn on the chalkboard.)

## Number of players: 2

Two teams consisting of the entire class can also be used.

## Object

Egyptian match provides practice in both the Egyptian numeration system and bilateral symmetry, or reflection on a line.

## How to play

One student or team uses Hindu-Arabic or Roman numerals; the other student or team uses Egyptian numerals. Team 1 selects its side of the design and writes a Hindu-Arabic or Roman numeral somewhere on the design. Team 2 must respond with the equivalent Egyptian numeral in the appropriate matching place on its side of the design. If either condition—equivalent Egyptian numeral or appropriate place—is not met, Team 1 wins. If Team 2 continues without error until the design is filled, it wins.

## After playing Egyptian match

Have students complete the Famous Dates activity.

**REFERENCE**

Weeks, John. *The Pyramids*. London: Cambridge University Press, 1974.

**Solution to Famous Dates**

1620—Pilgrims land at Plymouth
1492—Columbus discovers America
1776—Declaration of Independence signed
1976—American Revolution Bicentennial
1969—Neil A. Armstrong and Edwin E. Aldrin, Jr., land on the moon
1865—Civil War ends
1941—U.S. enters World War II

# EGYPTIAN NUMERATION SYSTEM

| Egyptian Numeral | Number Named | Meaning of Picture Symbol |
|---|---|---|
| \| | 1 | stroke |
| ∩ | 10 | cattle hobble or oxen yoke |
| ℓ | 100 | coil of rope |
| 𝄬 | 1 000 | lotus plant |
| ⌐ | 10 000 | bent finger |
| ⌒ | 100 000 | tadpole |
| ⩗ | 1 000 000 | a god with arms supporting the sky |

# EGYPTIAN MATCH

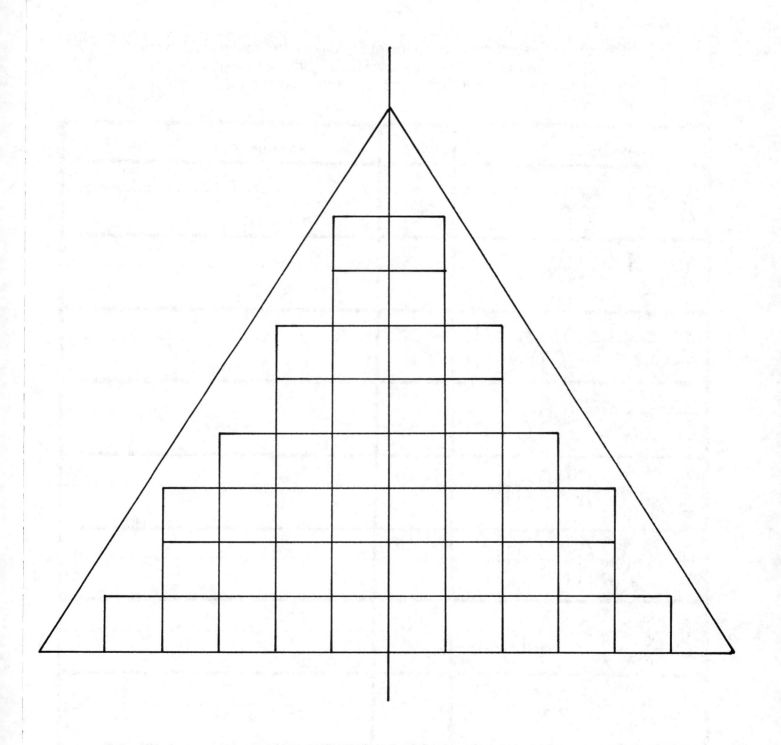

# FAMOUS DATES

Decipher the symbols of the ancient Egyptian numeration system to find the famous dates in history shown below. Then investigate further to find out what event in history occurred to make the date famous.

| Numeral | Date | Famous Event |
|---|---|---|
| 𓏏 𓍢𓍢𓍢 𓍢𓍢𓍢 ∩∩ | | |
| 𓏏 𓍢𓍢𓍢 𓍢 ∩∩∩∩∩∩∩∩ ‖ | | |
| 𓏏 𓍢𓍢𓍢𓍢 𓍢𓍢𓍢 ∩∩∩∩∩∩∩∩ ‖‖‖ | | |
| 𓏏 𓍢𓍢𓍢 𓍢𓍢𓍢 𓍢𓍢𓍢 ∩∩∩∩∩∩ ‖‖‖ | | |
| 𓏏 𓍢𓍢𓍢 𓍢𓍢𓍢 𓍢𓍢𓍢 ∩∩∩∩∩∩ ‖‖‖‖‖‖ | | |
| 𓏏 𓍢𓍢𓍢 𓍢𓍢𓍢 𓍢𓍢𓍢 ∩∩∩∩∩∩ ‖‖‖‖‖ | | |
| 𓏏 𓍢𓍢𓍢 𓍢𓍢𓍢 𓍢𓍢𓍢 ∩∩∩∩ | | |
| | | Your Birthday |

# SENET

## Background

Senet, the game of "passing," is an ancient Egyptian game. According to Piccione (1980), senet made its first known appearance in about 2686 B.C. The game was found in the tomb of an Egyptian named Hesy-re, who was an overseer of royal scribes. However, the game may be even older. Game boards of this type dating before this have been found.

From the numerous surviving game sets and the grave sites, Egyptologists know that senet was a popular game with all social classes; kings, nobles, and commoners played senet. Played until the beginning of the Christian era, senet evolved through the centuries from solely a recreational game to a game having both secular and religious forms. The secular form, played by two, continued to be recreational. The other form, played by one, was a ritual game with deeply religious significance.

The funerary senet was placed in tombs for the enjoyment of the dead in the next world. Religious thinkers of the time theorized that perhaps the fate of the dead depended on how well they played senet in the netherworld, or underworld. The dead played against their enemies or perhaps Fate itself to determine whether they would attain immortality. Kendall (1982) maintains that the living most likely played this game because it enabled them to fantasize about the adventures they felt awaited them after death. All this would lead, they hoped, to eternal life in heaven with Ra, the sun god.

The senet board has a path of thirty squares, or "houses." Originally, each player began with seven playing pieces positioned alternately on the first row and squares 11, 12, 13, and 14 of the second row. Square 15 was then the first square open for play.

Changes were made in the game during the Eighteenth Dynasty, 1570–1293 B.C. The number of playing pieces decreased to five. Gaming boxes were made, with senet on one side and a game called twenty squares on the reverse. A drawer in the middle of the box held playing pieces and casting sticks. An exquisite ebony, ivory, and gold game board of this type was found in King Tutankhamen's tomb.

The boards were decorated in a variety of ways. Early recreational boards were marked simply; later ritual boards pictured gods. Square 15 and the last five squares in row 3 seem to have had significance on all boards.

Although detailed rules of play are unknown, Egyptologists have provided information that gives a better understanding of the game. Consequently, senet can be played today.

## Materials

3 x 10 senet board
5 gold playing pieces
5 silver playing pieces
4 casting sticks—black on one side and white on the
    other. Half-round dowels or Popsicle sticks can be
    used.

# SENET (Continued)

*Number of players:* 2

## Object

Senet is a game of logic involving counting. Players attempt to pass the opponent and remove all their own playing pieces safely from the board. It is also a game of strategy. Players can develop strategies to block opponents and send them backward on the board. Occupying "good" squares and using blocking maneuvers can force an opponent to undesirable locations. Since moves are based on the toss of casting sticks, chance is also an element. Older students can determine why particular tosses are easy or difficult to obtain.

## How to play

Prepare the board by arranging the ten playing pieces on the top row, left to right, in the following order: silver, gold, silver, gold, and so forth. Playing pieces are moved from left to right along the top row, right to left along the middle row, and left to right along the bottom row:

| 1 | 2 | 3 | 4 | 5 | 6 | 7 | 8 | 9 | 10 |
|---|---|---|---|---|---|---|---|---|----|
| 20 | 19 | 18 | 17 | 16 | 15 | 14 | 13 | 12 | 11 |
| 21 | 22 | 23 | 24 | 25 | 26 | 27 | 28 | 29 | 30 |

Interpret the tosses of the casting sticks as follows:

One white side up equals 1
Two white sides equals 2
Three white sides equals 3
Four white sides equals 4
Four black sides equals 6

A toss of 1, 4, or 6 entitles the player to keep playing. After a toss of 2 or 3 and its move, play goes to the other player.

Interpret square 15 and the last five squares as follows:

• **Square 15:** "House of Repeating Life," a good square. If a playing piece lands here by exact count, a player is entitled to a free turn.

• **Square 26:** "House of Rejuvenation," a good square. If a playing piece lands here by exact count, the player is entitled to a free turn. From square 26, players try to cross over square 27, a definite pitfall.

• **Square 27:** "House of Humiliation," "House of Misfortune," or "Waters of Chaos." As the names imply, square 27 is a bad square. If a playing piece lands here by exact count, the player must make a choice: The playing piece must either be returned to square 15 or remain on square 27 *with no movement of any of the player's pieces* until a 4 is tossed. A 4 removes the piece from the board. If square 15 is occupied, the piece is placed on the empty square closest to square 15.

• **Square 28:** A good square, since the playing piece has successfully crossed over the "Waters of Chaos." A 3 must be tossed to remove the piece from the board.

• **Square 29:** A good square. A 2 must be tossed to remove the piece from the board.

• **Square 30:** A good square, since it is the square of Re-Horakhty (the sun god's name as he rises into the dawn). In the ritual game, removing the pieces from the board signifies the deceased's journey out of the netherworld and union with the sun god. A 1 must be tossed to remove the piece from the board.

Squares 15, 26, 28, 29, and 30 are "safe havens." Playing pieces on these squares are free from attack.

Gold plays first. Casting sticks are thrown to determine who plays gold, and the highest toss wins. Players take turns. A toss of the casting sticks can be used to move any *one* of a player's pieces. Note that if a 4 is tossed, the player cannot decide to move one piece one square and then a second piece three squares.

*Two playing pieces of the same color cannot occupy one square.* An opponent's playing piece is attacked by landing on the same square during a forward move. The opponent's playing piece is then moved back to the square left by the attacking piece.

When a forward move cannot be made, the turn must be used to move backward. Landing in reverse on a square occupied by an opponent moves the opponent to the forward square just vacated. If a move forward or backward is not possible, the turn is forfeited.

Playing pieces of the same color in two or three consecutive squares are safe from attack. Three playing pieces of the same color on three consecutive squares constitute a blockade. No other playing pieces can pass, but they can move backward if movement in that direction is possible.

To remove playing pieces from the board, players must make an *exact* toss of 4, 3, 2, or 1 when the pieces are in squares 27, 28, 29, or 30. If the number tossed is greater than the exact number needed to move a playing piece off the board, the player can either move another piece or move backward. If no forward or backward moves are possible, the player forfeits the turn. The player who gets all his or her playing pieces off the board first wins.

## REFERENCES

Grunfield, Frederic V., ed. *Games of the World.* New York: Holt, Rinehart & Winston, 1975.

Kendall, Timothy. "Games." In *Egypt's Golden Age: The Art of Living in the New Kingdom, 1558–1085 B.C.,* edited by Edward Brovarski, Susan Doll, and Rita Freed, pp. 263–70. Boston: Museum of Fine Arts, 1982.

Piccione, Peter A. "In Search of the Meaning of Senet." *Archaeology* 33 (July/August 1980): 55–58.

## Background

The Arabic word *mancala,* meaning "transferring," is the generic name for a set of similar board games played in a number of different regions of the world. Although the names, rules, and boards vary, the principles remain basically the same. Some games are played on boards having two rows of six holes; other, more complex games are played on boards having four rows of eight holes.

Mancala games have been known and played for thousands of years. In Egypt the boards have been discovered carved into the stone of the great pyramid of Cheops and the temples at Luxor and Karnak. The game spread to the rest of Africa and to Asia, the Philippines, the West Indies, and Surinam in South America as well.

Wari, also known as oware, is a two-row version of the game played by the Ashanti people of Ghana in West Africa. Located on the Gulf of Guinea, Ghana is predominantly agricultural; the biggest crop and major export is cacao (chocolate).

## Materials

1 egg carton
48 beans or pebbles

Remove the lid of the egg carton and cut it in half. Attach one half of the lid to each end of the lower portion of the carton.

**Number of Players:** 2

## Object of the Game

Wari is a game of logic—each player must consider the advantages and disadvantages of possible moves. For young students wari provides practice in counting. The object is to capture 25 beans, more than half the total. (If desired, the player having the most beans at the end of the game can be considered the winner.) Players must pay attention not only to their own side of the board but also to their opponent's side. Cups having one or two beans are vulnerable to capture. Players can defend their cups by moving their own beans so that one or two do not remain in a single cup. Or they can move their beans around the board to load the opponent's cups in such a way that the opponent's last bean will not land in the other player's vulnerable cups.

## How to play

To begin, four beans are placed in each egg cup. Players face each other with the board between them. A player's side of the board includes the six cups on his or her side and the lid, or pot, to the right. Movement is to the player's right—counterclockwise.

The **first player** removes four beans from any cup on his or her side and sows them *one by one* in the next four cups. On the first move it is possible, depending on which cup is emptied, to place beans in the opponent's cups. (Players do *not* sow beans in either of the pots; these are for captured beans only.) The **second player** removes all the beans from any cup on his or her side and distributes them similarly around the board.

Since cups can hold many beans, it is possible to go around the board more than once during a turn. As players go around the board, they skip the empty cup from which they last removed their beans and leave it empty for that turn.

Captures are made as follows:

1. If the last bean a player sows lands in a cup on the opponent's side that contains only one or two beans, that cup is then captured. *This last bean plus the opponent's bean(s) must total no more than three.*

2. When this condition is met, the player is also permitted to take beans from all cups preceding it if *(a)* each cup contains two or three beans, *(b)* the cups are on the opponent's side of the board, and *(c)* the cups are consecutive.

All captured beans as well as the last bean are placed in the player's pot.

No move is permitted that would enable the player to capture all the opponent's beans. Such a move would make it impossible for the opponent to play. When one side of the board is empty and it is the opposing player's turn, that player must move beans to the other side if possible so the game can continue. If, however, it is the player's side that is empty, then the game is over. The player who still has beans adds them to his or her own pot.

If a game continues on and on with only a few beans on the board, players may agree to stop. They then take the beans on their side of the board for their own pot, and the player with the most beans wins.

### REFERENCES

Grunfeld, Frederic V., ed. *Games of the World.* New York: Holt, Rinehart & Winston, 1975.

Zaslavsky, Claudia. *Africa Counts.* Boston: Prindle, Weber & Schmidt, 1973.

# KO-NO

Ko-no comes from China and Korea. *Ko-no* is the Korean name for games played on a diagram, and *k'i* is the Chinese name. The playing pieces are black and white stones.

### Materials

Game board, p. 9
2 black playing pieces
2 white playing pieces

### *Number of players:* 2

### Object

Ko-no is a game of logic. The object is to block the opponent so that he or she cannot move.

### How to play

One player uses the black pieces and the other uses the white ones. The black pieces, placed at *A* and *B*, are played first. The white pieces are placed at *C* and *D*. Players take turns by moving one playing piece along any segment to an unoccupied position. No jumps are allowed. The game is over when a player is blocked and cannot move either playing piece.

Discuss the winning and losing plays for ko-no. If both players play logically, what happens?

#### REFERENCE

Culin, Stewart. *Games of the Orient: Korea, China, Japan.* 1895. Reprint. Rutland, Vt.: Charles E. Tuttle Co., 1958.

# FIVE - FIELD KO-NO

In Korea, *ko-no* is the name for games played on diagrams. In addition to the ko-no game presented earlier, Koreans also played four-, five-, and six-field ko-no. Interestingly, five-field ko-no is played on a 4 x 4 diagram. Notice that a 4 x 4 diagram has *five* line segment intersections **across** (five rows) and five **down** (five columns). Playing pieces are moved along these lines.

The game's Korean name is *o-pat-ko-no.*

### Materials

Game board, p. 10
7 small black stones or playing pieces
7 small white stones or playing pieces

### *Number of players:* 2

### Object

To occupy the opponent's side of the board in the starting arrangement before the opponent occupies your side.

### How to play

Five-field ko-no is a strategy game. Arrange the playing pieces on the game board as shown. Black playing pieces move first. Moves are made along the line segments. A turn consists of moving one square forward, backward, or diagonally across a square, even though diagonals are not drawn on the board. Players take turns. Jumping is not allowed. The winner is the first player to get his or her playing pieces to the opposite side of the board in place of those of the opponent.

#### REFERENCE

Culin, Stewart. *Games of the Orient: Korea, China, Japan.* 1895. Reprint. Rutland, Vt.: Charles E. Tuttle Co., 1958.

# KO - NO

# FIVE-FIELD KO-NO

# TANGRAM

## Background

The tangram comes from China. Its origin is not known, and this adds an air of mystery to the puzzle. Many believe that the tangram must be, like other things from China, very old. Tangrams became known to the Western world during the nineteenth century. Sometimes people can be heard calling other geometric puzzles "tangrams," but the true Chinese tangram consists of seven special shapes: two large congruent triangles, one medium-sized triangle, two small congruent triangles, one square, and one parallelogram. These shapes will form—in addition to a square—a rectangle, a parallelogram, a trapezoid, and a triangle. Besides these geometric shapes, an infinite number of designs can also be formed.

## Materials

Scissors for each student
1 square of construction paper for each student, at least 15 cm × 15 cm

## Object

Constructing and manipulating the pieces, or "tans," of the tangram offer many opportunities for students to make geometrical discoveries concerning shapes, congruent triangles, and area. Students may create their own designs as well.

## Activity

Constructing the tangram pieces

1. Fold the square in figure 1 in half diagonally and cut on the fold. Two large congruent triangles result.
2. Fold one of the triangles in half and cut on the fold (fig. 1). Two triangles are formed. Save these two pieces.

3. Take the other large triangle ($\triangle ABC$) and fold $B$ to the midpoint of $\overline{AC}$ and cut off the triangle formed (fig. 2). Save this triangular piece.

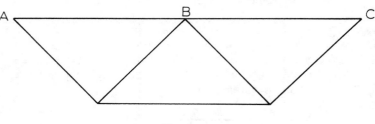

Fig. 2

4. The shape that remains is shown in figure 3. Fold the shape in half. $C$ and $Y$ are now on top of $A$ and $X$ (fig. 4).

Fig. 1

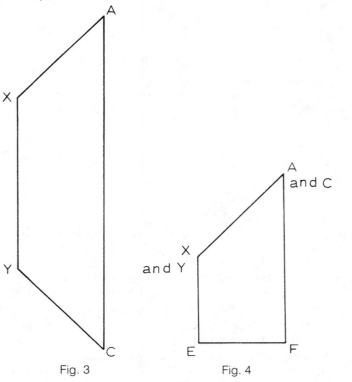

Fig. 3

Fig. 4

5. Unfold, and fold C to F (fig. 5).

Fig. 5

6. Unfold and fold X down to F (fig. 6).

7. The shape now has these three folds in it (fig. 7). Cut on the folds and save the four shapes. The puzzle pieces thus formed are two triangles, one square, and one parallelogram.

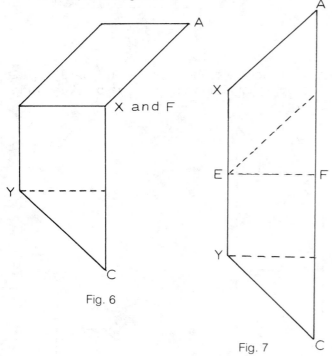

Fig. 6

Fig. 7

Have students consider the seven shapes found: five triangles, one square, and one parallelogram. They will find that the two large triangles and the two small triangles are congruent. Students may make other discoveries. The two small triangles will fit on top of the medium-sized triangle, the square, or the parallelogram.

**Activities**

1. Have students put the square together again (fig. 8).

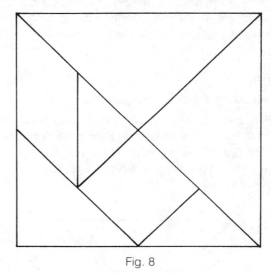

Fig. 8

2. Ask students to remove the two largest triangles from the solution. By positioning the two largest triangles around the remaining half of the solution, a rectangle, a parallelogram, a trapezoid, and a triangle can be formed.
3. Duplicate and distribute the cat puzzles that appear on the following pages. Younger students may match shapes, or tans, to each puzzle picture.
4. Trace the outlines of the cats. Older students may position the tans so that they fit within the outline.

**REFERENCE**

Read, Ronald C. *Tangrams—330 Puzzles*. New York: Dover Publications, 1965.

Cozy Cat

Curious Cat

Suspicious Cat

Dancing Cat

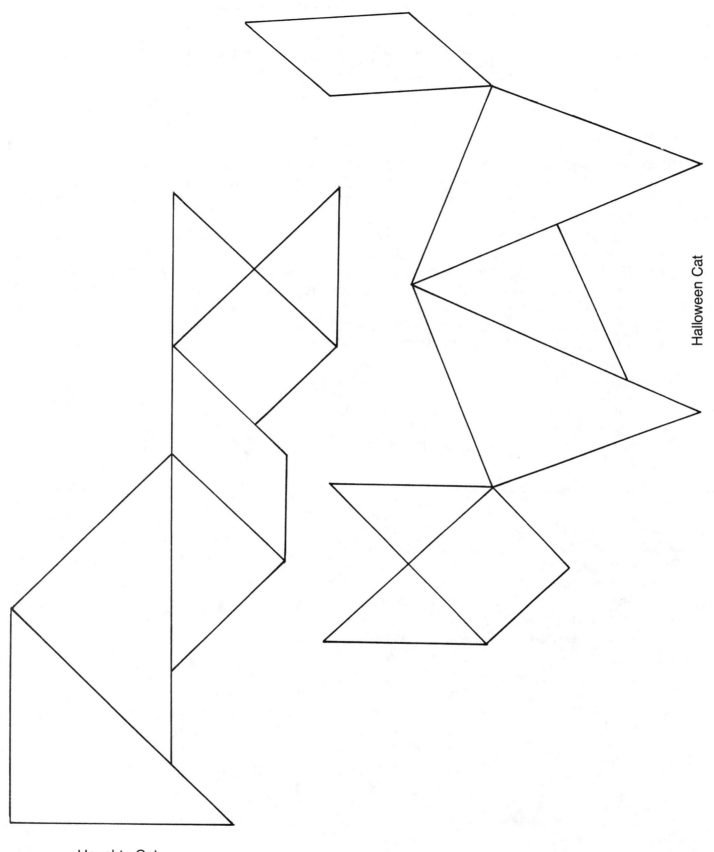

Halloween Cat

Haughty Cat

# ORIGAMI CUP

## Background

*Origami,* a Japanese word, means "the folding of paper." The origins of origami are unknown. In A.D. 538, a Buddhist priest brought methods of paper making to Japan from China and Korea. Logic dictates that methods of paper folding came in at the same time, although historians are unable to document this.

Since paper was expensive, ancient origami was used only for ceremonial occasions. The ancient ceremonial styles of origami were difficult, and to learn them people had to be taught by a specialist. Consequently, the ceremonial styles were not generally known. In the nineteenth century, after contacts with the West, the Japanese dropped the ancient origami styles in favor of modern works, those that are known today.

Origami is a handicraft enjoyed by both adults and children. Japanese mothers pass on paper-folding methods to their children.

## Materials

Paper squares 18 cm x 18 cm. Size of square and type of paper can vary. Since a drinking cup is being made, waxed paper can be used.

## Object

Students will construct a paper cup that can be used. The folding activity is geometrical. Geometric graphic aids and symbols are given in the instructions. This activity also provides practice in following instructions, which can be given orally, with a few illustrations on the chalkboard, or in written form.

1. Fold a square in half diagonally from corner to corner (fig. 1).

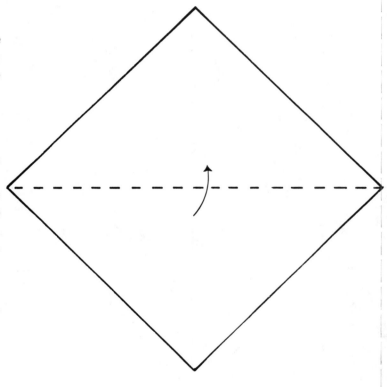

Fig. 1

**REFERENCE**

Honda, Isao. *The World of Origami*. 1965. Reprint. Tokyo: Japan Publications Trading Co., 1976.

(Continued)

## Origami (Continued)

2. Fold forward on $\overline{FG}$ (fig. 2), with *C* going to *D* (fig. 3).

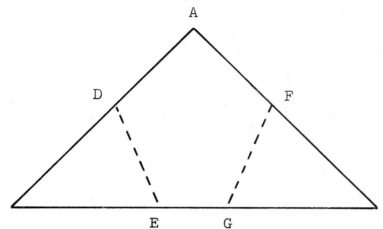

Fig. 2

3. Fold *backward*—away from your body—on $\overline{DE}$; *B* goes to the back of *F* (fig. 4).

Fig. 3

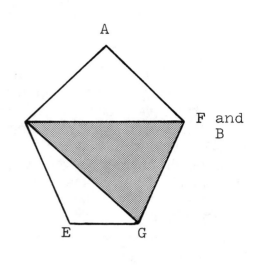

Fig. 4

4. Fold △*ADF* toward you and down into the paper slot (fig. 5).

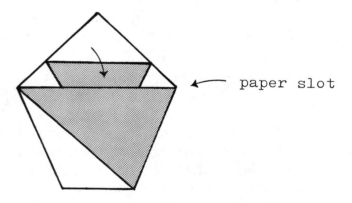

paper slot

Fig. 5

5. Fold the remaining top triangle down into the slot on the other side (fig. 6). The cup is now ready to use.

Fig. 6

# magic squares

## Background

A magic square is an arrangement of numbers in the form of a square so that when the numbers are added vertically, horizontally, or diagonally, the sum will be the same. Known from very ancient times, magic squares have been discovered in such widely divergent cultures as China, Egypt, India, and western Europe. It was believed that they had magic properties and perhaps were even connected with the stars. According to a Chinese myth, the magic square, called *lo-shu,* appeared on the back of a divine tortoise in the Yellow River about 2200 B.C.

## Materials

magic square grids, one or more for each student, pages 19 and 20

## Object

Magic squares provide practice in addition and subtraction. By following three rules, students can apply the pattern of movement to construct additional magic squares.

## Activity

To construct magic squares for odd numbers squared ($3^2$, $5^2$, $7^2$, . . . ), follow these three rules:

1. Position the numerals in consecutive order, beginning with 1. Place the numeral 1 in the top center cell.

2. Proceed diagonally upward and to the right from each small square. If you leave the large square at the top, drop to the bottom of the column. If you leave the large square at the side, go to the other end of the row.

3. If a number is a multiple of the number that is squared to get the total number of cells in the magic square, the next numeral is placed directly below. Figure 1 shows the pattern of movement for the construction of a ($3^2$) magic square.

Have students experiment with making magic squares using the set of natural numbers—the set of numbers beginning with 1 and having no largest member. This pattern of movement and square form works not only for consecutive numbers but for consecutive odd or even numbers as well. See figures 2 and 3.

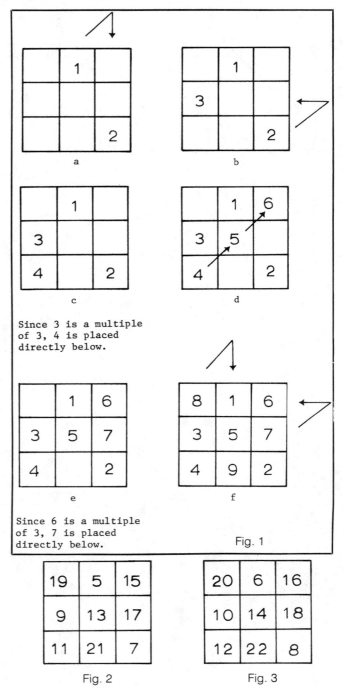

Since 3 is a multiple of 3, 4 is placed directly below.

Since 6 is a multiple of 3, 7 is placed directly below.

Fig. 1

| 19 | 5 | 15 |
|----|----|----|
| 9 | 13 | 17 |
| 11 | 21 | 7 |

Fig. 2

| 20 | 6 | 16 |
|----|----|----|
| 10 | 14 | 18 |
| 12 | 22 | 8 |

Fig. 3

18

# Magic Squares

| 17 |    | 1  | 8  | 15 |
|----|----|----|----|----|
|    | 5  | 7  | 14 |    |
| 4  | 6  | 13 |    | 22 |
| 10 | 12 | 19 |    | 3  |
| 11 | 18 |    | 2  |    |

Sum: _____

Sum: _____

| 34 |    | 2  | 16 | 30 |
|----|----|----|----|----|
| 46 |    | 14 | 28 |    |
| 8  | 12 |    | 40 | 44 |
| 20 | 24 |    |    | 6  |
| 22 |    | 50 | 4  | 18 |

# Magic Squares

**Make your own.**

| 30 | 39 | 1  | 10 | 19 | 29 | 37 |
|----|----|----|----|----|----|----|
|    | 7  | 9  |    | 27 |    |    |
| 46 | 8  |    | 26 |    | 36 | 4  |
| 5  | 14 | 16 | 25 | 34 | 44 | 12 |
| 13 | 15 | 24 | 33 | 42 | 3  | 20 |
| 21 | 23 |    | 43 | 2  | 11 |    |
| 31 | 40 | 49 |    |    |    |    |

इससे: _____

इससे: _____

इससे: _____

# LU - LU

GRADES 1 –6

## Background

Polynesian voyagers were the earliest known settlers of the Hawaiian Islands. The English explorer and navigator Captain James Cook discovered the islands and landed at Waimea, Kauai, on 20 January 1778. He named them the Sandwich Islands after John Montagu, the earl of Sandwich.

The islands are of volcanic origin. The Hawaiians played lu-lu with disks of vocanic stone about 2.5 cm in diameter. These disks served as stone dice called *u-lu*. The markings on the stone dice were painted red. The word *lu-lu* means "to shake."

## Materials

Four playing pieces marked on one side only as follows:

Playing pieces can be constructed from such materials as clay, shell, or wood. Buttons can also be used—masking tape can be affixed to them and marked with a felt-tip pen.

## Number of players: 2 or more

## Object

Lu-lu is an uncomplicated counting game.

## How to play

A player shakes the playing pieces in both hands and tosses them. A turn consists of two tosses:

- If all four playing pieces fall face up, the player scores ten and then tosses all pieces for a second time.
- If all four playing pieces do *not* fall face up, the player counts and scores the dots of those face up and then tosses *only* the face-down pieces a second time. The dots that show on the second toss are counted with those of the first toss.

The winner can be the player with the highest score in a single turn, or play can continue to an agreed-on score, such as 100.

**REFERENCE**

Culin, Stewart. "Hawaiian Games." *American Anthropologist,* n.s. 1 (April 1899): 201-47.

# KONANE

## Background

Konane (pronounced ko-nah-nay), the Hawaiian checkers game, was played by the Hawaiians before the first western explorers came to the islands. It is said that Kamehameha the Great, king of the island of Hawaii after 1790 and ruler of all the Hawaiian Islands from 1810 until his death, was a master of the game. Playing boards made from wood as well as stone have been recovered from ancient village sites.

## Materials

A game board, called a *papamu*. On old game boards, the number of positions varied from 64 to 260. For the classroom, mark 64 positions—8 rows of 8 dots—on a 30 cm × 30 cm piece of posterboard or cardboard.

The game pebbles, or *'ili'ili,* are as follows:
   32 white pebbles, or *'ili'ili ke'oke'o*
   32 black pebbles, or *'ili'ili 'ele'ele*
Pieces of water-worn coral were used for white, and black volcanic stones were used for black.

## Number of players: 2

## Object

The object of konane is to maneuver opponents to a position from which they can no longer move any of their playing pieces.

## How to play

Players cover the entire game board alternately with black and white playing pieces (fig. 1). To determine who begins, one player removes two adjacent playing pieces, one white and one black, from near the center of the board and conceals one in each hand. The other player chooses a hand, and the color of that playing piece is the choosing player's color. Black plays first. The two pieces are not returned to the papamu, or game board.

All moves are made by jumping. A jump consists of going over an opponent's playing

Fig. 1. Konane board layout

piece into a vacant place and removing the opponent's piece from the papamu. Multiple jumps on one turn are permitted.

Moves may be in a horizontal or vertical direction, never diagonally. Movement in more than one direction in a single turn is not permitted. It is permissible, however, for a player to decide not to take all the available jumps in a single turn.

With fewer and fewer playing pieces on the game board, fewer and fewer opportunities exist for jumping. The first player who is unable to jump over an opponent's playing piece loses the game.

### REFERENCES

Buck, Peter H. (Te Rangi Hiroa). *Arts and Crafts of Hawaii.* Section VIII, "Games and Recreation." Bernice P. Bishop Museum Special Publication no. 45, 1957. Reprinted in Separate Sections. Honolulu: Bishop Museum Press, 1964.
Mitchell, Donald D. *Hawaiian Games for Today.* Honolulu: Kamehameha Schools Press, 1975.

# PETROGLYPH

## Background

A *petroglyph* is a picture carved on a rock. Petroglyphs have been found in various parts of the world where early peoples lived. The petroglyph for this activity comes from Hawaii.

Little is known about the origin of the petroglyphs found in Hawaii. This is not surprising; historical evidence is usually not available for the origins of petroglyphs found in other parts of the world either. It is not unreasonable to assume that the Hawaiian petroglyphs were carved by the ancestors of the Hawaiians living today. The Hawaiians themselves, however, do not know who carved them or why.

The petroglyph for this activity represents a muscled figure. According to Cox and Stasack (1980), the portrayal of the human image as a triangular, muscled figure distinguished Hawaiian petroglyphs from petroglyphs found in other places. In addition to triangles, Hawaiian petroglyphs have circles, concentric circles, dots connected to curving lines, and U shapes. Interestingly, rectilinear forms were rarely used by the makers of the petroglyphs.

## Materials

petroglyph drawing
paper
pencil or marker

## Object of the Activity

The students draw a Hawaiian petroglyph of a muscled figure. In doing so, students are drawing pictures of triangles, simple closed figures having three sides. Older students can classify the triangles by their sides or angles.

## Activity

Ask students if they know what a petroglyph is. If no one knows, discuss how they can find out. After a definition has been obtained, introduce Hawaiian petroglyphs using the background information and petroglyph drawing provided.

Pose the following questions:

1. What shapes do you see in this petroglyph?

2. Do you think this petroglyph would—or would not—be easy to draw? Why?

Try it and see.

After drawing the petroglyph, older students can classify by sides the triangles they have drawn. That is, is a triangle isosceles, equilateral, or scalene? An isosceles triangle has two sides equal. An equilateral triangle has all three sides equal, whereas all three sides of a scalene triangle are unequal.

Older students can also classify the triangles they have drawn by angles. That is, is a triangle acute, right, or obtuse? An acute triangle has every angle measuring less than a right angle. A right triangle has one right angle. An obtuse triangle has one angle that is greater than a right angle.

### REFERENCE

Cox, J. Halley, and Edward Stasack. *Hawaiian Petroglyphs.* Bernice P. Bishop Museum Special Publication no. 60. Honolulu: Bishop Museum Press, 1970.

# PETROGLYPH

# TAPATAN

## Background

Tapatan is a Philippine game like three men's morris. The antiquity of this game can be easily documented, since it was one of the seven game boards cut into the roofing slabs of the temple at Karnak, Egypt ("Kurna" in some references). Consequently, the board can be dated from about 1400 to 1333 B.C., the period when the temple was constructed. Interestingly, the Chinese played the same game in about 500 B.C. during the time of Confucius. The Chinese called their game *luk tsut k'i,* "six man chess."

According to Bell (1979), the Latin poet Ovid (43 B.C.–A.D. 18) refers to the game in his *Ars amatoria* (the art of love). Most Roman boards were made of wood or stone, although boards for the wealthy were made of exotic materials.

Three men's morris was also popular in England around 1300. Evidence of this popularity remains in the great cathedrals, such as Canterbury and Westminster Abbey, where boards were cut into the cloister seats.

The Philippine game tapatan was collected in 1892 by Alexander R. Webb, then United States consul at Manila, for a collection he made for the United States National Museum. The overall hardwood board size measures 10.5 inches, or about 26 centimeters, square. The playing pieces for this board were six round pieces of wood, three white and three dark. Webb reported that many families liked and played tapatan; diagrams were often marked on the doorsteps or floors of native houses.

## Materials

tapatan board
3 white playing pieces
3 dark playing pieces

## Number of players: 2

## Object

A player's objective is to get three playing pieces of the same color in a line. Tapatan is played on a geometric field. Playing pieces can be placed or moved on nine points. The first player has three choices: the center, the middle of a side, or a corner. Note that the first player has the advantage and can force a win. After students discover which player can always win if no mistake is made, a spinner or die can be used to determine who plays first.

## How to Play

Dark plays first. The first player begins by placing one playing piece either in the center of the board where the line segments intersect or where any one of the line segments touches a side or corner of the square. The second player then places a playing piece on the board. Players take turns. After all six playing pieces are on the board, play continues with each player sliding a playing piece along a line to another position. Jumps are not allowed. The winner is the first player to get three playing pieces of the same color in a row, a column, or a diagonal.

### REFERENCES

Bell, R. C. *Board and Table Games from Many Civilizations.* 1960. Reprint. New York: Dover Publications, 1979.

Culin, Stewart. "Philippine Games." *American Anthropologist,* n.s. 2 (October–December, 1900): 643–56.

McConville, Robert. *The History of Board Games.* Palo Alto, Calif.: Creative Publications, 1974.

# TAPATAN

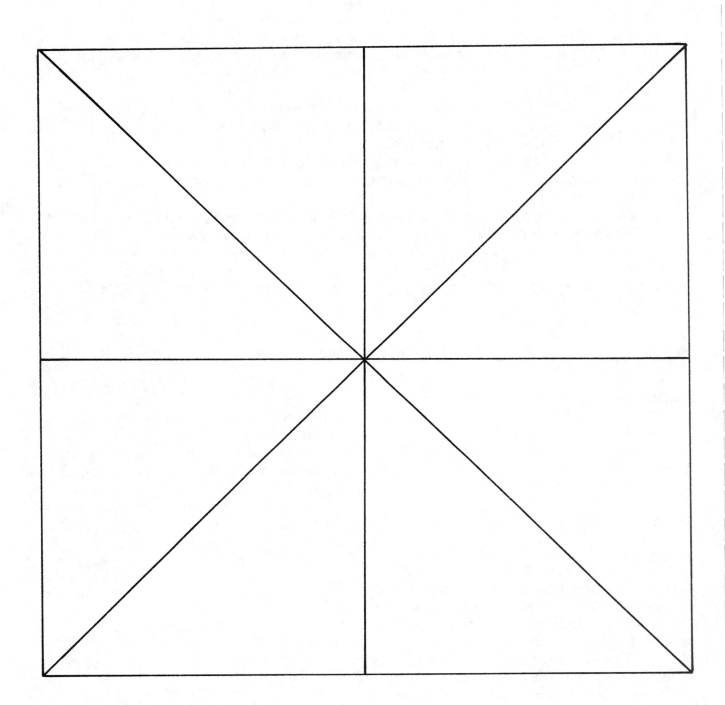

# Nine Men's Morris

## Background

Nine men's morris, also called *merels, mill,* and *morelles,* is an ancient game for two. This game can be traced back to about 1400 B.C., for the game board was one of the seven cut into the roofing slabs of the temple in Karnak, Egypt ("Kurna" in some references). Similar diagrams have been found in Sri Lanka, formerly Ceylon. There, two diagrams dating from A.D. 9–21 were carved into the flight of steps on the lower part of the hill at Mihintale. Evidence shows that the game was known to the inhabitants of the first city of Troy as well as to the Vikings. A Bronze Age burial site in County Wicklow, Ireland, also revealed a morris board carved in stone.

According to Bell (1979), nine men's morris reached the zenith of its popularity in Europe during the fourteenth century. Even Shakespeare mentioned this game in 1598 in *A Midsummer Night's Dream.* In act 2, scene 1, lines 98–100, he wrote:

> The nine men's morris is fill'd up with mud,
> And the quaint mazes in the wanton green
> For lack of tread are undistinguishable.

## Materials

nine men's morris board
9 red playing pieces
9 green playing pieces

## Number of players: 2

## Object

Nine men's morris is a reasoning and strategy game played on three concentric squares. Each player attempts either to block an opponent's playing pieces so that a move is not possible or to reduce the opponent to two playing pieces.

## How to play

Players decide who plays first. The game begins with a clear board. Players take turns placing playing pieces, one at a time, at any angle or intersection of the figure. After all the playing pieces are on the board, play continues with pieces moved one at a time along a line to an adjacent empty position. Jumps are not allowed.

If a player succeeds in getting a mill—a row of three playing pieces on any line vertically or horizontally—he or she can remove one of the opponent's playing pieces from the board. However, if an opponent's playing piece is in a mill, it cannot be removed from the board.

New mills can be made by opening existing mills. That is, a playing piece can be moved to form a mill and then returned on the next turn. Each time a mill is made, the player is entitled to remove one of the opponent's playing pieces.

If a player has only three playing pieces left on the board and they form a mill, then on his or her turn the player must break the mill. This may result in the player losing the game.

A player wins when the opponent has only two playing pieces left on the board or is unable to move.

### REFERENCES

Bell, R. C. *Board and Table Games from Many Civilizations.* 1960. Reprint. New York: Dover Publications, 1979.

Shakespeare, William. *The Comedies of Shakespeare.* Introduction by A. C. Swinburne. London: Humphrey Milford, Oxford University Press, n.d.

# Nine Men's Morris

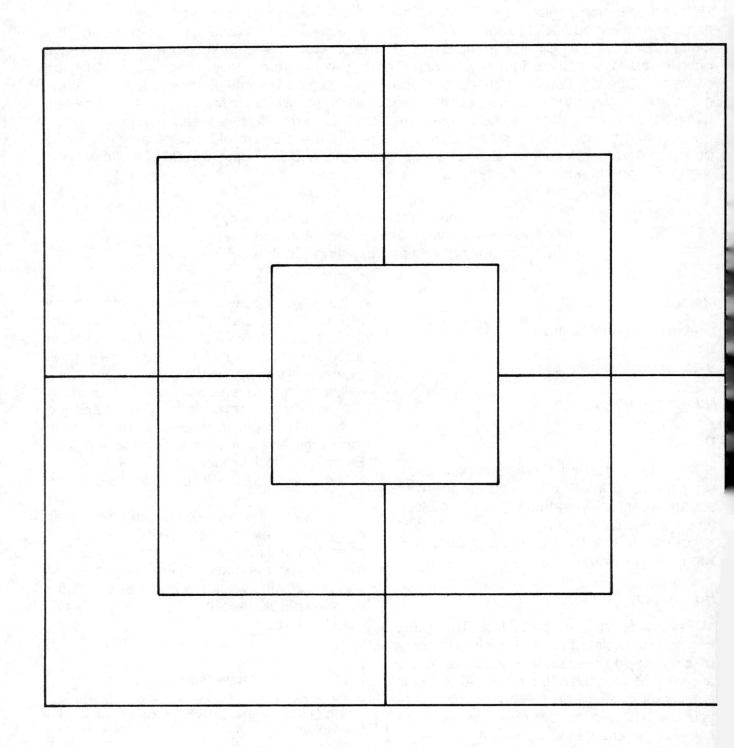

# JULEKURV

## Background

Norway is a mountainous country extending into the Arctic Circle. Most of Norway borders Sweden, but in the north it borders Finland and the USSR. The coastland of Norway is indented by many beautiful fjords—narrow inlets of the sea between high banks or cliffs.

The julekurv is a Norwegian Christmas basket. The basket is usually made from a flexible high-gloss paper that comes in many colors. Norwegians hang these baskets filled with sweets on their Christmas trees.

## Materials

2 strips of different colored construction
  paper 9 cm × 30 cm
construction-paper strip for handle
pencil
ruler having centimeter and millimeter divisions
scissors
glue

## Object

Students measure in centimeters and use decimal fractions.

## Activity

Cut 2 strips of construction paper 9 cm × 30 cm and perform the following steps on *each* strip:

1. Fold in half—9 cm × 15 cm.
2. Divide the strip into four equal parts along the fold. Each division will be approximately 2.2 cm.
3. Draw three 9.5-cm line segments extending from the three dividing points on the fold. See figure 1.
4. Cut along the three line segments.
5. Cut a semicircle opposite the fold.
6. To begin weaving the two strips X and Y together, position the pieces as in figure 2.
7. Place A inside 1, 2 inside A, A inside 3, and 4 inside A. One row is woven. Slide this row to the top of the three line-segment cuts in strip X.

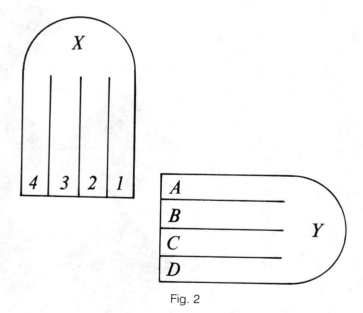

Fig. 1

Fig. 2

8. Place 1 inside B, B inside 2, 3 inside B, and B inside 4.
9. Place C inside 1, 2 inside C, C inside 3, and 4 inside C.
10. Place 1 inside D, D inside 2, 3 inside D, and D inside 4.
11. Attach handle.

29

# JULEKURV

# ERIN GO BRAGH

## Background

The national emblem of Ireland is the shamrock. The leaves of the shamrock plant have three leaflets. Legend tells us that Saint Patrick, who lived from about 385 to 461, used the shamrock to illustrate the doctrine of the Trinity. A shamrock, either real or artificial, is traditionally worn on 17 March, Saint Patrick's Day. Both Irish and non-Irish people can be seen wearing a shamrock on this day.

## Materials

Erin Go Bragh page
ruler
green crayon or marking pen
scissors

## Object

By connecting a given sequence of points with straight line segments, students construct shamrocks. The shamrock design gives the illusion of curves, although no curves are drawn.

## Activity

Distribute materials. If desired, students can number the remaining sides that are not numbered. As the students work from the center of the design (9 to 1, 8 to 2, 7 to 3, and so on), discuss the acute angles that have sides in proportion. An equal number of marks are designated on both sides of the angles. After they have drawn the line segments, have students draw a stem. Students may color the shamrock, cut it out, and wear it on Saint Patrick's Day.

Solution

# ERIN GO BRAGH

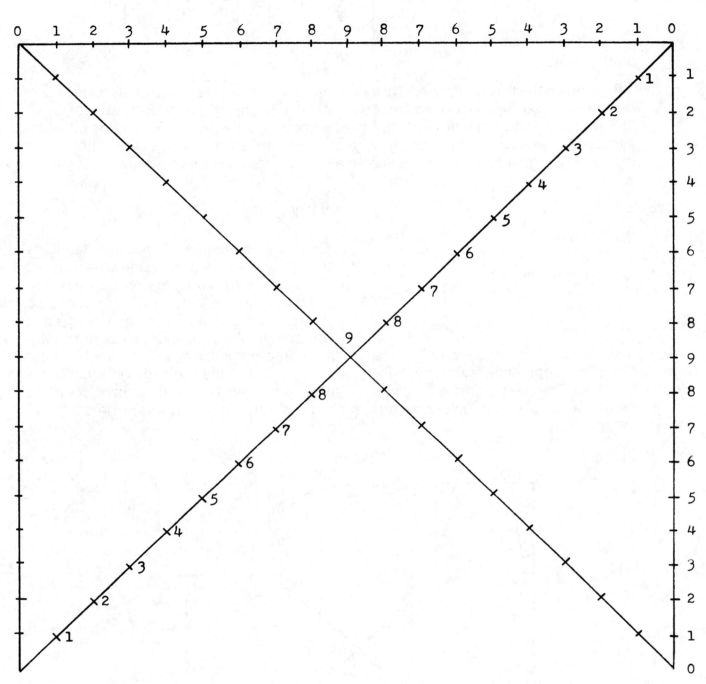

Begin at the center and draw line segments from

| | | |
|---|---|---|
| 9 to 1 | 6 to 4 | 3 to 7 |
| 8 to 2 | 5 to 5 | 2 to 8 |
| 7 to 3 | 4 to 6 | 1 to 9 |

# Roman Numerals

## Background

Tradition says that Rome was founded by Romulus in 753 B.C. The Romans were ruled by the Etruscans until about 500 B.C., when the Romans overthrew the Etruscans and established a republic controlled by the patrician class. This republic survived for four centuries. The Roman Empire is dated from 27 B.C. and the rule of Augustus, the grandnephew of Julius Caesar, to the defeat and death in A.D. 476 of Romulus Augustulus, the last Emperor of the West.

Mathematics did not flower during the time of the Romans. This is surprising, since the Greeks had a great thirst for mathematical knowledge and the Romans often imitated the Greeks. According to Cajori (1980), the mathematics of the Romans probably came in part from the ancient Etruscans.

Their system is unusual in that the principle of subtraction is used. If a symbol of lesser value is placed before a symbol of greater value, then the lesser value is subtracted from the greater value. That is, XL means 50 − 10, or 40. But if the lesser value is placed after the greater value, the two values are added. That is, LX means 50 + 10, or 60. When a horizontal bar is placed over a symbol, the value is multiplied by 1000. The $\overline{M}$ means 1000 × 1000, or 1 000 000. The Roman system of numeration did not have a symbol for zero.

The Roman system of numeration is still in use today. Roman numerals are seen on clocks and watches and in book chapters, outlines, and formal inscriptions.

### Roman Numerals

| | | |
|---|---|---|
| I | = | 1 |
| V | = | 5 |
| X | = | 10 |
| L | = | 50 |
| C | = | 100 |
| D | = | 500 |
| M | = | 1000 |

## Materials

Roman Numerals activity page

## Object

The Roman Numerals activity page provides practice in renaming Hindu-Arabic numerals as Roman numerals. Also, one Roman numeral must be renamed as a Hindu-Arabic numeral. This activity can be used in conjunction with history, each date representing a famous historical event. (Other dates may be added or substituted as desired.)

## Activity

Distribute the activity page and review with students the values of the seven Roman numerals given on the left. Review the principles of subtraction and addition used in the Roman system. Then ask students to rename each famous date as a Roman numeral.

### Solution

| | |
|---|---|
| MDCCLXXVI | — The signing of the Declaration of Independence |
| MCMLXIX | — Neil A. Armstrong and Edwin E. Aldrin, Jr., land on the moon |
| MCMXXVII | — Charles Lindbergh makes the first nonstop flight from New York to Paris |
| MCDXCII | — Columbus discovers America |
| 1976 | — American Revolution Bicentennial |
| MDXIII | — Ponce de León discovers Florida and Balboa discovers the Pacific Ocean |
| MCMLXXXI | — Space Shuttle *Columbia* launched on its first orbital flight |

### REFERENCE

Cajori, Florian. *A History of Mathematics.* 3d ed. 1893. Reprint. New York: Chelsea Publishing Co., 1980.

# Roman Numerals

Fill in the blank spaces.

| Roman Numeral | Value of Numeral | Date | Roman Numeral |
|:---:|:---:|:---:|:---:|
| I | | 1776 | |
| V | | 1969 | |
| X | | 1927 | |
| L | | 1492 | |
| C | | | MCMLXXVI |
| D | | 1513 | |
| M | | 1981 | |

# DREIDEL

## Background

A dreidel is a four-sided top. Hebrew letters are on the four sides of the dreidel:

 nun     gimmel     hay     shin

These symbols are the initial letters of the Hebrew words *nes gadol hayah sham,* "a great miracle happened there," referring to an event that took place over two thousand years ago. After Judas Maccabee and only a few soldiers defeated the army of a cruel king who attempted to make the Jewish people worship idols, the Jews lighted the temple lamp. There was only oil in the lamp for a few hours and yet the lamp burned for eight days and eight nights. This is why the Festival of Lights, Hanukkah, is celebrated for eight days.

## Materials

1 dreidel (Inexpensive dreidels can be purchased—contact a local synagogue for information.)

10 pieces of "Hanukkah gelt" for each player (Dried beans, peanuts, candy, small stones, or pennies can be used.)

## Number of players: 2 or more

## Object

The dreidel game is a game of chance providing practice in adding, subtracting, and using the common fraction ½.

## How to play

*Before beginning play,* each player places an agreed-on number of beans in the center of the table or in a pot. A turn consists of a player spinning the dreidel and following these directions:

1. If *nun* turns up, *nothing* is taken from the pot. The next player then spins.

2. If *gimmel* turns up, the player takes *all* the beans in the pot. When the pot is empty, each player again places the earlier specified amount in the pot. The next player then spins.

3. If *hay* turns up, the player takes *half* the beans in the pot. For an odd number, the player takes half plus one. The next player then spins.

4. If *shin* turns up, the player *adds* the agreed-on number of beans to the pot. The next player then spins.

The game is over either when one player has won all the beans or after a designated number of rounds. One round is completed after each player has spun the dreidel once.

## Variation

Give each letter a numerical value and play to get the highest score, such as 1000. Or the winner can be the player with the highest score after a designated number of rounds.

| | | | |
|---|---|---|---|
| nun | 50 | hay | 3 |
| gimmel | 5 | shin | 300 |

## After playing

Have older students experiment as follows:

1. If you spin a dreidel 100 times, how many times would you expect the dreidel to turn up nun, gimmel, hay, and shin? Record what you expect.

| Letters | Expected Outcomes |
|---|---|
| ‬ב‬ (nun) | |
| ‫א‬ (gimmel) | |
| ‫ה‬ (hay) | |
| ‫ש‬ (shin) | |
| TOTAL | 100 |

Note that the outcomes are equally likely.

2. Try this experiment and see what happens. Compare your results with what you expected. Tell what happened.

3. Play the game again. Keep a record of each player's spins. Do the results surprise you?

## REFERENCE

Goodman, Philip. *The Hanukkah Anthology.* Philadelphia: Jewish Publication Society of America, 1976.

# TESSELLATIONS

## Background

Our word *tessellate* comes from the Latin *tessella,* diminutive of *tessera,* "a small square block." The noun *tessellation* comes to us from the Latin *tessellare,* "to pave with tiles."

Tessellations have been used for many centuries. The Romans used small tiles to make mosaics for their floors. In 1941, excavations on the south side of the Cologne cathedral in Germany revealed the remains of a second-century Roman house and its now famous floor, known as the mosaic of Dionysos. The mosaic measures 74 m². This floor mosaic is composed of approximately two million small blocks of limestone, marble, slate, pottery, and blue or green glass pastes.

Muslim artisans have demonstrated exceptional skill in juxtaposing shapes in coherent geometrical patterns. The Islamic religion forbids the use of the human form. Arabic buildings are often decorated with colorfully tiled facades. The interiors may also contain exquisite decorative mosaics and carvings.

When students are studying tessellations, the works of the Dutch artist M. C. Escher (1898–1972) may be of interest. Escher created fascinating pattern designs.

## Materials

construction-paper shapes, such as squares, triangles, rectangles, and hexagons
tessellation activity pages
tessellation background grid
ruler
compass
colored pencils or markers (optional)

## Object

Students will complete pattern designs—tessellations. Students should also be encouraged to develop a pattern design of their own on the tessellation background grid.

## Activity

Study the geometric shapes that fit together without any spaces or gaps. Students can demonstrate with the construction-paper shapes which shapes will fit together. Discuss the background and meaning of the word *tessellation.* Ask students if they have noticed in their surroundings any pattern designs composed of geometric shapes. Flooring, chicken wire, and checkerboards are a few examples.

Distribute a tessellation activity page and consider with the students the shape formed in relationship to the square. Interestingly, a more exotic design often has as its basis a simple shape. Have students complete the design.

**REFERENCE**

Mold, Josephine. *Tessellations.* Cambridge: At the University Press, 1969.

# Tessellation Activity Page

# Tessellation Activity Page

# Tessellation Activity Page

# tessellation background grid

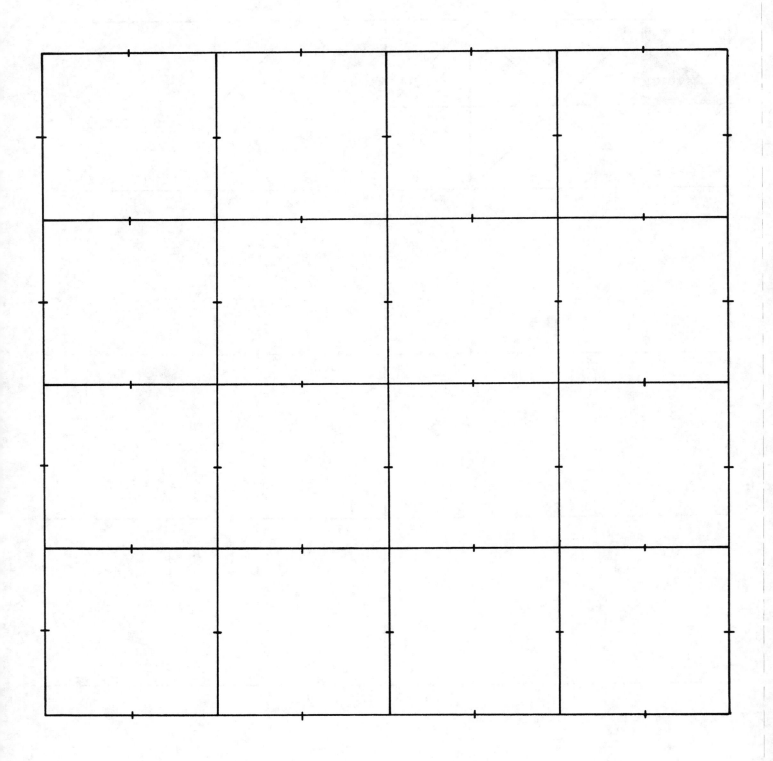

# QUIPU

## Background

The Incas used *quipus,* pronounced kē′ pōōs, for keeping records and sending messages. Information such as population and resources was recorded on the quipu. The quipu was a counting device based on the decimal system. Since the Incas left no written record, the method of deciphering is unknown.

The quipu consisted of a top cord or rope to which smaller cords or strings of different colors were attached. Authorities believe that a string's color, position, thickness, and the number of knots tied in it were all significant.

## Materials

1 piece of rope or cord
strings or yarn of different colors (Use one string for each place value being studied.)

## Object

Students represent numbers in base ten.

## Activity

Have the students make a quipu for base ten. The number of strings attached will depend on the place values the students are studying. Assign place values to the different colored strings. Tie knots in the strings to represent different numbers.

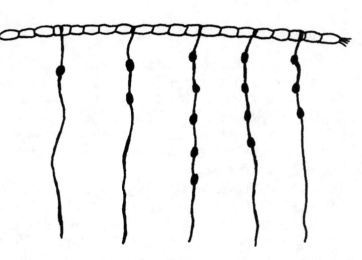

Fig. 1. A quipu representing 12 543.

# INCA BIRD DESIGN

## Background

The center of the Inca empire was Cuzco, Peru. The Incas founded Cuzco during the eleventh century. Their empire extended over 3000 kilometers, north to south, along the western coast of South America. The land of the Incas also reached inland and included the Andes Mountains.

The word *Inca* may refer to the emperor, who ruled by divine right, but it is commonly used by historians to refer to the people or to the empire itself. The Inca empire of six million people was conquered in 1533 by Francisco Pizarro and his force of fewer than two hundred Spanish soldiers.

The Incas are well known for their network of roads, building constructions, pottery, weaving, and uses of copper, bronze, tin, silver, and gold. The Inca bird of this activity was a woven design.

## Materials

Inca bird design
transfer grid

## Object

Using a grid system of squares, the student transfers and enlarges the design. If placed on the large grid, the design could be *reduced* by transferring it to the small grid.

The mathematical basis for scale drawings is similarity. The ratio of similarity is the scale given for a drawing.

The small grid and the large grid are similar figures; that is, they have the same shape. The scale is doubled, since the length of the side of the large square is double the length of the side of the small square. But the area of the large shape is four times that of the small shape. Note that four 1 cm × 1 cm shapes will fit on one 2 cm × 2 cm shape.

## Activity

Distribute the Inca bird design and the transfer grid. Discuss the uses of scale in our lives—for example, diagrams, maps, or photographs that are either enlarged or reduced. Through the centuries, travelers have reached their destinations using scale to determine distance and direction. Scale enables the traveler to look at a map and measure the distance between points. Perhaps the most familiar kind of scale is the "inches-to-the-mile" scale. It is an amazing fact that one inch of paper can represent one mile, or 63 360 inches, of our planet. Also, one centimeter of paper can represent ten kilometers, or 1 000 000 centimeters, of our planet.

After the students have enlarged the design, ask about the scale and area of the new shape.

Enlarge the design by drawing it on the transfer grid. Use the following key: 1 small unit → 1 large unit.

### REFERENCE

Glubok, Shirley. *The Art of Ancient Peru.* New York: Harper & Row, Publishers, 1966.

# INCA   BIRD   DESIGN

Enlarge the design by drawing it on the transfer grid.
Use the following key: 1 small unit → 1 large unit.

# Transfer Grid

# THE MAYA

## Background

The ancient Mayan culture flourished in Central America and southern Mexico between approximately A.D. 250 and 900. The Mayan numeration system included the use of zero, place value, addition, and the repetition of symbols. The system was essentially vigesimal, or base twenty. Figure 1 shows an exception to the vigesimal system: 18 uinals, *not 20*, make 1 tun.

The Maya were fascinated by the passage of time and often expressed time intervals. They had two sophisticated calendars: the *haab*, a 365-day year with fifty-two-year cycles, and the *tzolkin*, a sacred or ceremonial year of 260 days. The tzolkin had thirteen 20-day months. The corrected calendar year with eighteen 20-day months plus one 5-day month approximates the astronomically determined year with greater accuracy than the Gregorian calendar does.

The Maya wrote their numerals vertically. That is, reading from bottom to top the values are as follows: 1's, 20's, 360's, 7 200's, and so on. Three symbols were used in counting: a dot represented 1, a horizontal bar represented 5, and a shell represented 0. (See figs. 2 and 3.)

| | |
|---|---|
| 20 kins, or days | = 1 uinal, or 20 days |
| 18 uinals | = 1 tun, or 360 days |
| 20 tuns | = 1 katun, or 7 200 days |
| 20 katuns | = 1 cycle, or 144 000 days |
| 20 cycles | = 1 great cycle, or 2 880 000 days |

Fig. 1

## Object

Students are given an opportunity to work with a numeration system other than base ten by constructing a calendar using the Mayan numerical symbols. The Mayan numeration system, essentially a base of twenty, had the position of the values written vertically.

## Activity

Discuss the Maya and their numeration system. Distribute the Maya Numerical Symbols sheet and point out the placement of the values vertically. Twenty, for example, is written as one 20 and zero 1's. One hundred is written as five 20's and zero 1's. Have students practice writing the Mayan symbols for 1 through 31.

Distribute the calendar grid. Have the students select a month and create a Mayan calendar of their own using the Mayan numerical symbols.

## Materials

Maya Numerical Symbols sheet
calendar grid

$$1 \times 360 = 360$$
$$3 \times 20 = 60$$
$$7 \times 1 = \underline{\phantom{00}7}$$
$$427$$

Fig. 2

$$8 \times 360 = 2880$$
$$11 \times 20 = 220$$
$$14 \times 1 = \underline{\phantom{0}14}$$
$$3114$$

Fig. 3

### REFERENCES

Cajori, Florian. *Notations in Elementary Mathematics.* A History of Mathematical Notations, vol. 1. LaSalle, Ill.: Open Court Publishing Co., 1928.

Smith, David Eugene. *Special Topics of Elementary Mathematics.* History of Mathematics, vol. 2, 1925. Reprint. New York: Dover Publications, 1958.

# MAYA NUMERICAL SYMBOLS

| Hindu-Arabic | Maya | Hindu-Arabic | Maya |
|:---:|:---:|:---:|:---:|
| 0 | ⬭ | 7 | •• |
| 1 | • | 8 | ••• |
| 2 | •• | 9 | •••• |
| 3 | ••• | 10 | ═ |
| 4 | •••• | 15 | ≡ |
| 5 | — | 20 | •⬭ |
| 6 | •— | 100 | ⬭ |

## My Mayan Calendar for _____

| SUN. | MON. | TUES. | WED. | THUR. | FRI. | SAT. |
|---|---|---|---|---|---|---|
|  |  |  |  |  |  |  |
|  |  |  |  |  |  |  |
|  |  |  |  |  |  |  |
|  |  |  |  |  |  |  |
|  |  |  |  |  |  |  |

# PATOLLI

## Background

The Aztecs were Indian people who dominated central Mexico when the Spaniards under Cortés arrived in 1519. The Aztecs played patolli on a mat game board in the shape of a cross. Within the cross, line segments formed squares, or houses. The patolli game board resembles the *pachisi* board. Beans called *patolli* were used as one-sided dice.

The earliest Spanish account of patolli was written by Francisco Lopez de Gomara, a Spanish chronicler, in 1552, thirty-one years after the conquest, and reveals that Emperor Montezuma sometimes watched as the game was played.

Before the game, players would bring fire, toss incense into the flames, and make offerings of food to Macuilxochitl ("Five Flowers"), the god of the dice. During the game, players would call on not only Macuilxochitl but also Ometochtli, the god of gambling, to give them good points on the dice.

The game was played enthusiastically, and betting was heavy. After the conquest, the Aztecs stopped playing the game, apparently because of their superstitions.

Since too few of the original rules of the game are known for play, the rules below are suggested. Players can add penalties to the game if desired.

## Materials

game board
5 lima beans
Mark a single dot on one side of each bean with a marking pen.
If there are two players, each needs six playing pieces of the same color; if three players, four playing pieces each; and if four players, three pieces each. Playing pieces could be small colored stones or construction paper shapes.

## Number of players: 2 to 4

## Object

Patolli is a game of chance where moves are determined by the toss of five beans.

## How to play

Each player has his or her own starting place at a different arm of the game board— A, B, C, or D. Players take turns. For the toss, the beans are held in both hands. Determine the order of play by tossing the beans and counting the dots facing up. If players tie, they toss the beans again. The highest scorer plays first and tosses the beans again for the first turn.

The first playing piece can be brought onto the game board with any toss. The remaining pieces can be brought onto the game board only when a 1 is tossed. Count the tosses as follows:

1 dot up, move forward 1 space
2 dots up, move forward 2 spaces
3 dots up, move forward 3 spaces
4 dots up, move forward 4 spaces
5 dots up, move forward 10 spaces

The first player to move all his or her playing pieces completely around the patolli game board is the winner. To leave the board, the player must have the exact toss for the number of spaces left.

## REFERENCE

Tylor, E. G. "On the Game of Patolli in Ancient Mexico, and Its Probably Asiatic Origin." *Journal of the Anthropological Institute of Great Britain and Ireland* 8 (1879): 116–31.

47

# PATOLLI

# AZTEC CALENDAR

## Background

The Aztecs lived in the Valley of Mexico and surrounding areas from about A.D. 1200 until 1521, when they were conquered by the Spanish under the leadership of Hernán Cortés. (From this conquest came the beginnings of the Mestizo people and the Mexico of today.) The capital city of the Aztec empire was Tenochtitlán, pronounced tay-nohch-TEE-tlahn, which is now Mexico City.

One of the most magnificent expressions of Aztec art is the Sun Stone, or Aztec calendar (fig. 1). The calendar attests to the Aztecs' knowledge of astronomy and mathematics. Hieroglyphs for their days, months, and suns (or cosmic cycles) are pictured on the calendar.

The carved stone is 3.6 meters (12 feet) in diameter and has a mass of twenty-four metric tons. It took fifty-two years to make the calendar, from 1427–1479. Most authorities believe only stone tools were used. This calendar is superior to calendars used by Julius Caesar or Cleopatra and is 103 years older than the Gregorian calendar created by Pope Gregory XIII in sixteenth-century Rome. The Gregorian calendar is used worldwide today.

Originally the Aztec calendar was placed atop the main temple in Tenochtitlán. Today, Mexico City's cathedral stands on the site. The Aztec calendar faced south in a vertical position and was painted in brilliant red, blue, yellow, and white.

Tenochtitlán fell in 1521 to the Spaniards, who destroyed the temple and buried the stone. The stone was lost for over 250 years; in December 1790 it was found quite by accident during repair work on the cathedral. Today, the Aztec calendar is in the National Museum of Anthropology in Mexico City.

The face of Tontiuh, the Aztec sun god, is in the center circle of the stone. Around the face are four squares called *Nahui-Ollin,* or "Four Movement." According to Aztec legend, these squares represented the different ways that the four previous suns (or worlds) had come to an end: first by wild animals, then by wind, by fire, and by floods. The Aztecs believed they were living in the fifth and last world.

Continuing outward, the next concentric circle shows twenty squares, each naming one of the twenty different days of the Aztec month. Clockwise these days are as follows:

1. Snake—Coatl
2. Lizard—Cuetzpallin
3. House—Calli
4. Wind—Ehecatl
5. Crocodile—Cipactli
6. Flower—Xochitl
7. Rain—Quiahuitl
8. Flint—Tecpatl
9. Movement—Ollin
10. Vulture—Cozcacuauhtli
11. Eagle—Cuauhtli
12. Jaguar—Ocelotl
13. Cane—Acatl
14. Herb—Malinalli
15. Monkey—Ozomatli
16. Hairless Dog—Itzquintli
17. Water—Atl
18. Rabbit—Tochtli
19. Deer—Mazatl
20. Skull—Miquiztli

(Continued)

Fig. 1

## Aztec Calendar (Continued)

The Aztec year consisted of eighteen months, each having 20 days. Each month was given a specific name. This arrangement took care of 360 days (18 × 20), to which five dots were added inside the circle. These dots, known as *Nemontemi,* were sacrificial days.

The next concentric circle is composed of square sections with five dots in each section, probably representing weeks of five days. Next are eight angles dividing the stone in eight parts. These represent the sun's rays placed according to the cardinal points.

On the lower portion of the stone, two enormous snakes encircle the stone and face each other. Their bodies are divided into sections containing the symbols for flames, elephant-like trunks, and jaguar-like forelegs. It is believed that these sections are also records of fifty-two year cycles. A square is carved at the top of the calendar between the tails of the snakes. Inside the square the date 13 *Acatl* (or *Cane*) is carved. This corresponds to 1479, the year the calendar was finished.

Eight equally spaced holes appear on the very edge of the calendar. The Aztecs placed horizontal sticks here, and the shadows of the sticks would fall on the figures of the calendar; thus the stone also served as a sundial.

### Materials

Aztec Numerical Symbols sheet
calendar grid

### Object

Students will construct a calendar using Aztec numerical symbols. The Aztec numeration system had a base of twenty. Units of one were represented by a dot. A diamond stood for 10; a flag, 20; a feather, 400; and a bag with tassels, 8000. The Aztecs applied repetitive and additive principles: for example, 423 is 400 + 20 + 3, or a feather, a flag, and three dots.

### Activity

Discuss the Aztecs and their calendar. Distribute the Aztec Numerical Symbols sheet and discuss the repetitive and additive principles of the numeration system. Have students practice writing the symbols 1–31 in the Aztec system.

Distribute the calendar grid. Have the students select a month and create an Aztec calendar of their own using the Aztec numerical symbols; check that students are beginning the calendar with the first day in the correct position on the grid. The name of the month should be written at the top of the calendar. Students may also decorate the top of the calendar with flowers—the Aztecs loved flowers.

### REFERENCES

Dávila, Francisco González. *Ancient Cultures of Mexico.* Mexico City: Museo Nacional, 1961.

Ragghianti, Carlo Ludovico, and Licia Ragghianti Collobi. *National Museum of Anthropology Mexico City.* Milan: Newsweek & Arnoldo Mondadori Editore, 1970.

"Stone of the Fifth Sun." *National Geographic* 158 (December 1980): 757–59.

# AZTEC NUMERICAL SYMBOLS

| Hindu-Arabic | Aztec | Hindu-Arabic | Aztec |
|:---:|:---:|:---:|:---:|
| 1 | • | 9 | ⦂⦂⦂ |
| 2 | •• | 10 | ◇ |
| 3 | ••• | 20 | ⎏ |
| 4 | •• •• | 100 | ⫲ |
| 5 | •• •• | 200 | ⫲⫲ |
| 6 | ••• ••• | 300 | ↓ |
| 7 | ••• •• ••• | 400 | ↓ |
| 8 | ••• ••• ••• | 8000 | (symbol) |

## My Aztec Calendar for _____

| SUN. | MON. | TUES. | WED. | THUR. | FRI. | SAT. |
|---|---|---|---|---|---|---|
| | | | | | | |
| | | | | | | |
| | | | | | | |
| | | | | | | |
| | | | | | | |

# RAIN BIRD

## Background

The Pueblo Indians were—and are—prolific pottery makers. The Zuni Indians are one of the Pueblo Indian tribes. A design appearing frequently on Zuni water jars is known as the rain bird. How the term *"Rain Bird"* came into existence is unknown; however, Mera (1970) believes that the design's ancestry may be traced to the beginnings of pottery decoration in the Southwest. There have been many variations of the design over the last three centuries.

The Zuni Indians live in western New Mexico. They farm using irrigation. They are noted not only for their pottery but also for basketry, jewelry, weaving, and ceremonial dances.

## Materials

paper
pencil or crayon
ruler (optional)
protractor (optional)

## Object

The rain bird motif can be used with young students to study the triangle. Older students can construct the different kinds of triangles: equilateral, isosceles, right, and scalene.

## Activity

Have the student draw a picture of a triangle. Next the hook-like head, legs, and three or four tail feathers are added. Discuss with students why they think the rain bird de-sign was popular with the Zuni Indians. Discuss the rainfall in the geographical area where they live.

**REFERENCE**

Mera, H. P. *Pueblo Designs.* New York: Dover Publications, 1970.

# PAIUTE WALNUT SHELL GAME

## Background

This game comes from the Paiute Indians of Pyramid Lake in northwestern Nevada. The Paiutes of this area belong to the group designated as the Northern Paiutes, as opposed to the Southern Paiutes. The Northern Paiutes were scattered over central and eastern California, western Nevada, and eastern Oregon. The Southern Paiutes, who were also known as the Digger Indians because they subsisted on roots, lived in northwestern Arizona, southeastern California, southern Nevada, and southern Utah.

## Materials

8 walnut shell halves, filled with pitch and powdered charcoal (In the classroom, clay can be used. Small red and white beads and pieces of abalone shell are inlaid.)
basket tray
small sticks or twigs

**Number of players:** 2 or more

## Object

The Paiute walnut shell game provides practice in one-to-one correspondence and counting to reach an agreed-on sum.

## How to play

Place the walnut shell halves in the basket tray and toss. If three or five of the walnut shells land with the flat side up, the player scores one point. No point is scored for any other combination. Each player keeps count of his or her score with small twigs. Players take turns. The winner is the first player to reach an agreed-on sum.

### REFERENCE

Culin, Stewart. *Games of the North American Indians.* 1907. Reprint. New York: Dover Publications, 1975.

# TEPEES

## Background

The Cheyenne, Blackfoot, and Sioux American Indian tribes lived on the Great Plains and were hunters. They lived in conical structures called tepees (or tipis). The tepee was made with a frame of poles and was usually covered with buffalo hides. A three-pole foundation was used in the central plains, and a four-pole foundation in the northwest plains. These foundation poles were tied together near their tops and provided support for additional poles. Mobility was a significant feature of the tepee. The diameter of tepees ranged from about three meters for a hunting tepee to about nine meters for a camp tepee. They were sometimes beautifully painted with ceremonial designs.

## Materials

brown construction paper
scissors

## Object

Folding paper-chain tepees demonstrates a repeating geometrical pattern. By the use of translation, the student can slide the tepee pattern to the next position.

## Activity

Folding paper-chain tepees out of brown construction paper.

1. To make equal divisions, fold paper in half, in half again, and in half again.
2. Open and refold like a fan with accordian folds.
3. Using figure 1 as a guide, draw the tepee on the first folded section, and then slide the pattern on to the next position. Continue to the other end of the paper, being sure to position the middle of the tepee (dotted lines in the figure) on each fold.

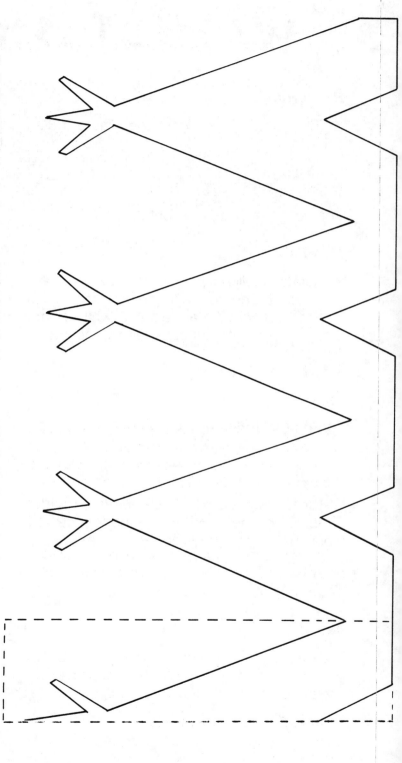

Fig. 1

# EYE - DAZZLER

## Background

The Navajo (or Navaho) constitute the largest Indian tribe in the United States. During the seventeenth century, these Native Americans lived in the region between the San Juan and Little Colorado rivers in northeastern Arizona. Because their language is related to Athabascan, spoken by their northern neighbors, authorities believe that the Navajo migrated here from farther north.

At one time they were predatory—raiding not only their neighbors the Pueblo Indians but also the Spanish and Mexican settlements in New Mexico. However, with the introduction of sheep by the Spanish in the seventeenth century, the Navajo became pastoral nomadic people. In the winter, they lived in hogans—round or hexagonal earth-covered structures; in the summer, they lived in brush structures. The door of a Navajo home always faces eastward to the rising sun.

Navajo women have become well known for their skill in weaving. With the establishment of trading posts in Navajo country, it was not long before blankets became trade items. Since the white settlers did not wear the blankets as the Navajo did, they put the blankets to such other uses as bed and floor coverings. Traders purchased blankets by the pound and blankets sold in this manner became known as "pound blankets."

## Materials

4-squares-to-an-inch or centimeter graph
    paper
eye-dazzler design page
ruler

## Object

The eye-dazzler design provides practice in reproducing a given pattern. The eye-dazzler blanket style was popular with the Navajo in the 1880s. Note the diamond pattern. Eye-dazzler blankets exploded with vibrant colors. This "blanket" may be colored with red, yellow, and orange.

Students may reproduce a design or create one of their own. See the design on centimeter paper (fig. 1).

## Activity

Distribute the eye-dazzler design page and graph paper. Have students reproduce the pattern or create their own "eye-dazzler."

### REFERENCE

Kahlenberg, Mary Hunt, and Anthony Berlant. *The Navajo Blanket*. n.p.: Praeger Publishers in association with the Los Angeles County Museum of Art, 1976.

Fig. 1

Navajo Eye-Dazzler Design

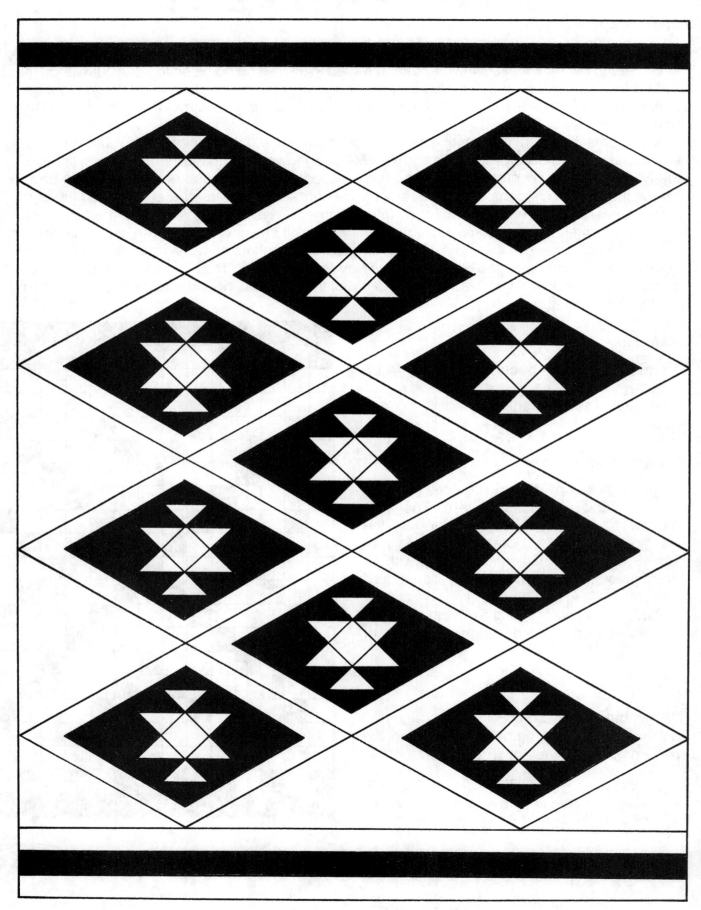

# NAVAJO FRY BREAD

## Background

The Navajo Indians call themselves the *Dîné̂h,* which means "the people of the earth." Moving with their sheep to find good pasturage, the Navajos did not have electricity in the remote areas where they lived. This fact is reflected in the fry bread recipe—they did not have refrigerators to keep milk fresh and therefore used powdered milk.

## Materials

ingredients given below
electric skillet

## Object

To make Navajo fry bread, students must measure ingredients and follow a set of step-by-step instructions. Note that Navajo cooks use pinches and handfuls. This recipe, given to the writer at Cameron, Arizona, in the Navajo Nation, has been "translated" into both customary and metric measurements.

## Activity

Ingredients
2 cups flour (250 g)
2 teaspoons baking powder (10 g)
½ teaspoon salt (2.5 g)
½ cup powdered milk (125 mL)
1 cup warm water

Sometimes a small amount of shortening is added to the ingredients; however, this is usually left out. Mix together the flour, baking powder, salt, and powdered milk. Add warm water to form dough. Knead the dough until it is soft but not sticky. Cover with a cloth and let stand for about 2 hours. Shape the dough into balls about 2″ (5 cm) across. Flatten by patting the balls into circles about 8″ (20 cm) in diameter. Place the circles in hot shortening about 3 cm deep. Fry at 375 °F (190 °C) until golden brown on one side, then turn and fry the other side until golden brown.

A sandpainting of a rainbow figure, a Navajo deity.

# Hopi Rain Cloud

## Background

In northeastern Arizona, three rugged promontories project from the south side of Black Mesa, a great circular upland having a diameter of about ninety-six kilometers. To the Hopi, Black Mesa is *Tuuwanasavi,* "Center of the Universe." It is where the migrations of the Hopi clans met and settled about a thousand years ago. These migrations had taken the Hopi in the four directions to the ends of the land. Following the instructions of Maasaw, guardian spirit of the Hopi Fourth World, some clans turned north while others turned south. Retracing their routes, they then turned east and west and returned again.

The rain cloud symbol is frequently used by the Hopi. It may be placed over altars and on roof slabs, on kachinas (dolls), pottery wares, and even gaming implements. This is not surprising, since Hopi existence at one time was dependent on rainfall. Farming is as important to the Hopi today as it was to their ancestors. Farming in an arid region, they grow beans, corn, melons, squash, and cotton. Corn represents the *mother of life.*

## Materials

activity page or other paper
compass
ruler

## Object

The Hopi Rain Cloud activity involves the construction of semicircles, or half circles. The prefix *semi* signifies half; a semicircle is half a circle.

## Activity

Have students first look at a circle with diameter $\overline{AB}$. If the points A and B are the endpoints of the diameter, each of the two arcs—above and below—is called a semicircle. The arc degree measurement of a semicircle is 180. This is written as $m(\widehat{AB}) = 180$.

*Constructing the rain cloud symbol*

1. Use the activity page or draw a line segment in about the middle of the paper. Select a length that is divisible by four.

2. Mark and label points A, B, and C at the following locations (see fig. 1):
   A—one-fourth of line segment from left endpoint
   B—one-fourth of line segment from right endpoint
   C—midpoint of the line segment and above it at a distance that is one-fourth of the line segment

3. Place the point of the compass at A and twist the compass to construct a semicircle above the line. Repeat at B and C (see fig. 2).

4. Draw short vertical line segments for rain (see fig. 2).

5. Draw "lightning snakes" from the clouds if desired (see fig. 3).

**REFERENCE**

Fewkes, J. Walter. "A Few Summer Ceremonials at the Tusayan Pueblos." In *A Journal of American Ethnology and Archaeology,* vol. 2, pp. 1–160, edited by J. Walter Fewkes. New York: Houghton, Mifflin & Co., 1892.

# Hopi Rain Cloud

C
.

A                                    B

Fig. 1

## Directions

1. Place the point of your compass at *A* and twist the compass to construct a semicircle above the line. Repeat at *B* and *C* (see fig. 2). These semicircles are the rain clouds.

2. Draw short vertical line segments for rain (see fig. 2).

3. Draw "lightning snakes" from the clouds if desired (see fig. 3).

Fig. 2

Fig. 3

# Pottery Designs

## Background

Hopi pottery making began about fifteen hundred years ago in the Four Corners area where Arizona, Utah, Colorado, and New Mexico meet. From the beginning, the Hopi decorated their pottery with geometric designs. The following designs were found on pottery ware recovered from the ruins of Sikyatki, an ancient pueblo. The destruction of Sikyatki took place before 1540. The cause of the destruction is unknown. Sikyatki was excavated in 1895 by the Smithsonian Institution under the direction of J. Walter Fewkes. Fewkes referred to this activity as the "Expedition to Arizona in 1895."

The elaborately decorated pottery that combined life forms and geometric designs is known today as Sikyatki Polychrome. Fewkes considered this pottery the "most highly developed" of Hopi ceramics.

## Materials

Pottery Designs activity page
ruler

## Object

The Pottery Designs activity page provides practice in completing repeating patterns. These patterns were bands encircling the exterior of Sikyatki food bowls. Mathematically, the patterns are obtained by a combination of translation—simply sliding to the next position—and reflection on an axis.

## Activity

### 1. Pottery Designs, activity 1

As each band is studied, discuss with students the geometrical shapes and line segments of the design. Ask if translation, reflection, or a combination of both will be used to obtain the completed pattern.

Cardboard patterns can be cut out using the two-band shapes. The top band has alternating right triangles that meet at their right angles, and the lower band has two isosceles triangles meeting at their top vertexes. Students then will be able to slide and flip the patterns to check their answers.

### 2. Pottery Designs, activity 2

The composition of this exterior band is more complex than the bands in activity 1. The overall form is rectangular; however, three right triangles, another rectangle, and five sets of vertical parallel bands are within the rectangular band.

To complete the top band, have each student flip the pattern mentally first. The completed design is a Sikyatki food-bowl decoration.

For the lower band, pose the following question: "What would the band decoration look like if the Hopi pottery maker had decided *not* to flip the pattern?"

**REFERENCE**

Fewkes, Jesse Walter. *Designs on Prehistoric Hopi Pottery.* 1898. Reprint. New York: Dover Publications, 1973.

Solution

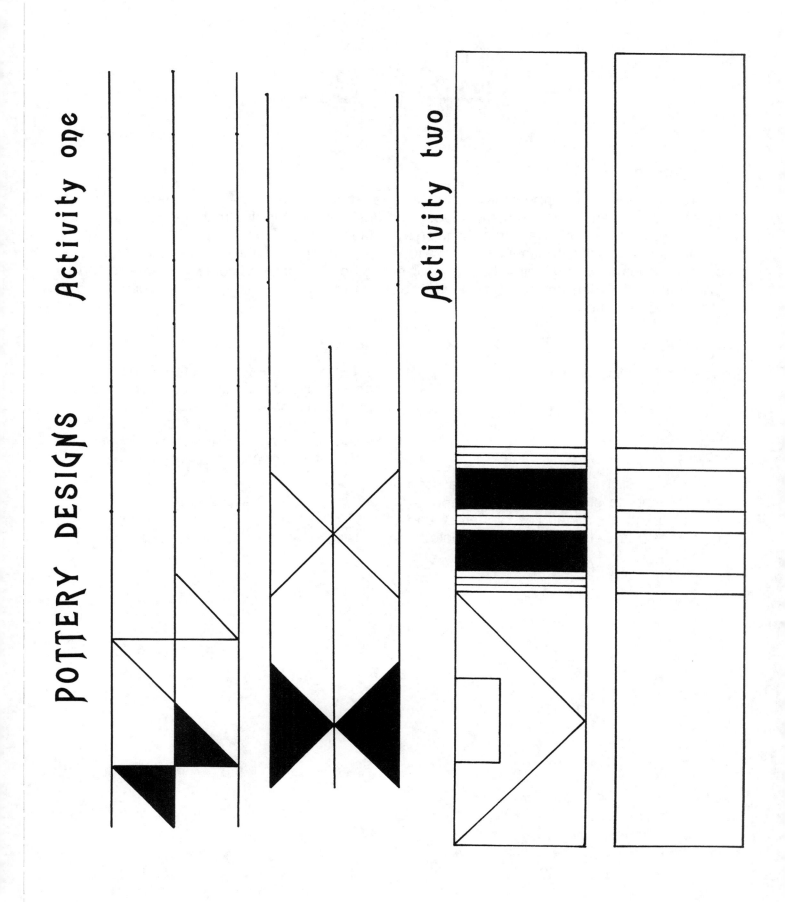

POTTERY DESIGNS

Activity one

Activity two

# Hopi Bird

## Background

The Hopi Indians of northeastern Arizona are well known for their beautiful pottery ware, which is decorated lavishly with geometric shapes and designs. The bird figure that follows was found on a piece of prehistoric ware from ruins on the Little Colorado River.

Hopi pottery making continues to flourish today. Hopi potters use the coil method to make their pottery. Ropes of clay are spiraled, one on top of another. The pot is smoothed, dried, and coated with watery clay. Then the pot is polished with a smooth stone. The design is carved or painted on the pot before firing.

## Materials

Hopi Bird activity page

## Object

The Hopi Bird activity page can be used with students to study geometric shapes—circles, triangles, rectangles, squares, and trapezoids. Parallel lines can also be noted.

## Activity

Distribute the activity page. Explain to students that this is a figure found on a piece of prehistoric pottery made by the Hopi. Ask what living thing they think this figure represented and why. Ask students to point out the different geometric shapes they see in the Hopi Bird.

### REFERENCE

Fewkes, Jesse Walter. *Designs on Prehistoric Hopi Pottery*. 1898. Reprint. New York: Dover Publications, 1973.

# Hopi Bird

# Sky Window

## Background

The Hopi sky god, Co-tuk-i-nung-wa, is represented by a five-pointed star. The star symbolizes the deity's eye, through which he watches all things. The eye of Co-tuk-i-nung-wa is also known as the sky window. Through the sky window, the sky god gives all things vitality.

## Materials

paper
compass
protractor

Sky Window

## Object

The sky window can be used as a compass-and-protractor activity. Students inscribe a pentagon in a circle and construct a pentagram, or star.

The sky window here is drawn from a Polacca Polychrome jar, 1820–1860, on display in the Heard Museum's exhibition, "America's Great Lost Expedition: The Thomas Keam Collection of Hopi Pottery from the Second Hemenway Expedition, 1890–1894." The Heard Museum is in Phoenix, Arizona.

## Activity

*Inscribing a pentagon inside a circle (fig. 1)*

1. Draw a circle having a radius of 5–8 cm with a compass.
2. Label the radius $\overline{AB}$.
3. Divide the circle (360°) into five equal parts of 72° each and mark these points.
4. Draw connecting lines from the center to the points, constructing "angles" of 72° (solid lines in figure).
5. Label each new radius drawn, $\overline{AC}$, $\overline{AD}$, $\overline{AE}$, and $\overline{AF}$.
6. Draw line segments (dotted lines in figure) from B to C, C to D, D to E, E to F, and F to B to form a pentagon.

*Constructing a pentagram or star (fig. 2).*

Knowledge of the pentagon will enable students to construct the pentagram.

1. Divide the circle into five equal parts and mark these dividing points as before.

2. Draw line segments from B to D, B to E, C to E, C to F, and D to F.
3. Erase the circle and the line segments inside the pentagram.

### REFERENCE

The Heard Museum. *America's Great Lost Expedition: The Thomas Keam Collection of Hopi Pottery from the Second Hemenway Expedition, 1890–1894.* Phoenix: Heard Museum, 1980.

Fig. 1

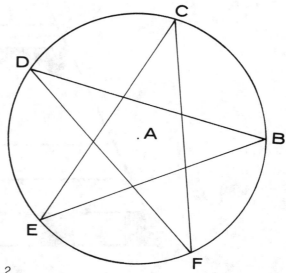

Fig. 2

# Sun Kachina Mask

## Background

Hopi kachinas are believed to be supernatural beings embodying the spirits of living things or of dead ancestors. The Hopi men who don masks and impersonate the supernatural beings in the kivas (underground structures) and plazas of the villages are also called kachinas. The Hopis believe that when a man impersonates a kachina he acquires the kachina's spirit.

The Sun Kachina, also called Tawa Kachina, represents the spirit of the Sun God. The mask is a yucca basket having a forehead that is red on the left side and yellow on the right side. The remaining face is painted green. Surrounding the face are eagle feathers.

## Materials

Sun Kachina Mask activity page
mirrors (optional)

## Object

The Sun Kachina Mask activity provides practice in mirror geometry—that is, this is an activity in symmetry with respect to a straight line. The right half of the mask is to be a mirror image of the other half, an example of bilateral symmetry. *Bi* means "two," and *lateral* refers to the sides. To say it simply, the two sides will match.

## Activity

Distribute the activity page. Discuss the meaning of kachinas and ask students why they think the sun commands such importance. Consider where the Hopi live—in a semiarid region of northeastern Arizona.

Have students place a mirror perpendicular to the line in the middle of the mask. This line divides the mask into two congruent parts. The missing part on the right side can be obtained from the part on the left. After the students have studied the effect of the mirror, ask them to complete the mask. Clearly this is an activity for older students and not a beginning activity for young students. Even so, time spent experimenting with mirrors to obtain mirror images would be well spent.

# Sun Kachina Mask

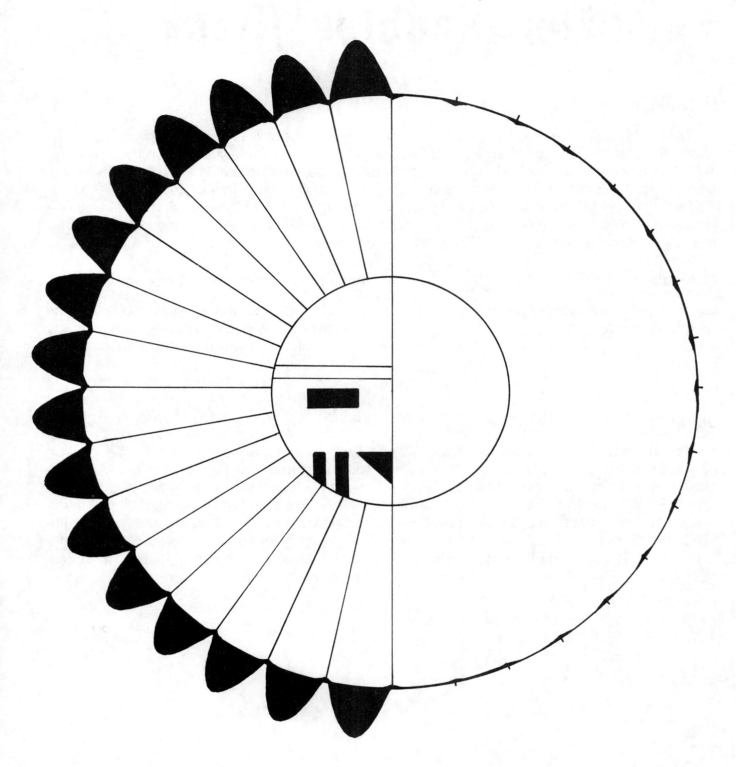

# Additional Activity Pages

## Background

*Sikyatki Butterfly Design*

For the Hopi, the butterfly is a symbol of fruitfulness. Butterfly or moth figures appear on the pottery unearthed in 1895 at the ruins of the ancient pueblo Sikyatki. Anthropologist J. Walter Fewkes identified five typical butterfly or moth figures found on Sikyatki pottery. An element common to all was a triangular body.

Without question the most beautiful figure of all is the butterfly design on the famous "butterfly vase." The figure is repeated six times, which probably represents a relationship to the cardinal points—north, west, south, east, above (or zenith), and below (or nadir).

The six butterfly heads face the opening of the vase. All the butterflies do not have the same head markings but alternate in the following manner: A butterfly having a cross on the circular head with a dot in each section is alternated with a butterfly having only dots on the head and no cross. The former represents a female butterfly and the latter represents a male butterfly. North corresponds to male, west to female, south to male, east to female, above to male, and below to female.

*Sun Design*

The sun, or *Tawa,* is the life giver, the father. The Hopis asked Tawa to grant them good harvests.

## Materials

Sikyatki Butterfly Design activity page
Sun Design activity page
mirrors (optional)

## Object

These half designs provide additional practice using the concept of bilateral symmetry, or symmetry about a line.

## Activity

Distribute an activity page and let students make their own unhurried investigations with mirrors. Ask if the design is bilaterally symmetric to the line. Have students complete the design.

### REFERENCE

Fewkes, Jesse Walter. *Designs on Prehistoric Hopi Pottery.* 1898. Reprint. New York: Dover Publications, 1973.

# Sikyatki Butterfly Design

# Sun Design

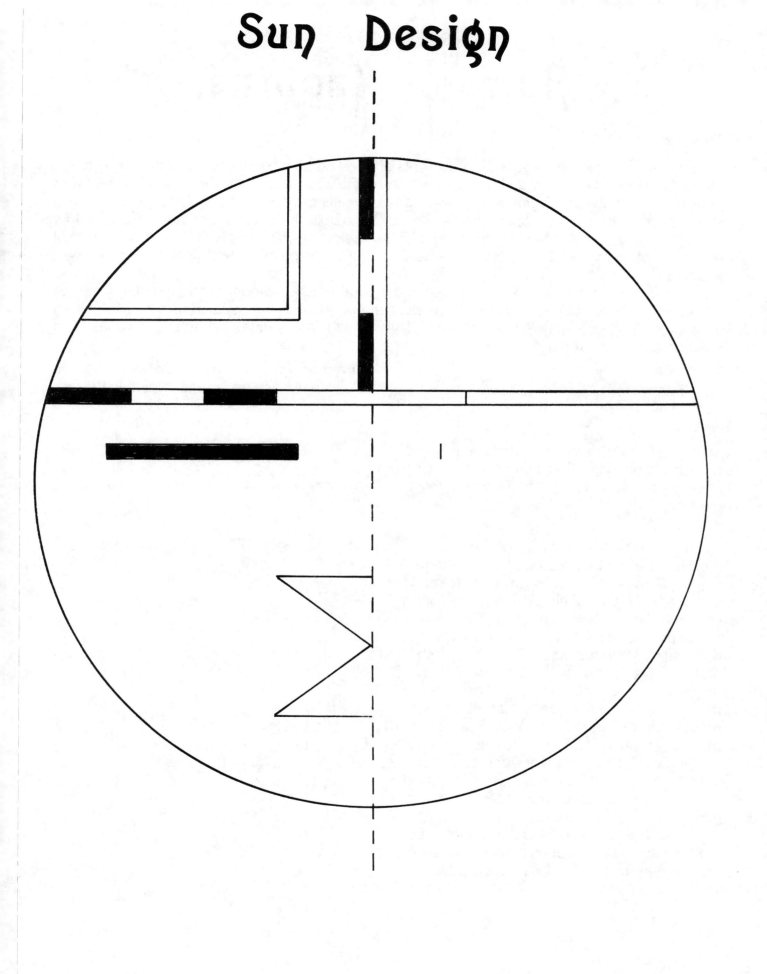

# Hemis Kachina

## Background

Hopi men carve kachina dolls from dry roots of dead cottonwood trees, which they find along the banks of the Little Colorado River or along one of the washes near the three Hopi mesas. After cutting the cottonwood to the length desired and whittling the piece of wood into the desired shape of the kachina, the Hopi craftsman smooths the doll with a wood rasp and sands it with a piece of sandstone. A nose, horns, ears, and the headdress, or *tableta,* are added with small wooden pegs or glue. The kachina doll is then ready for painting and final adornment with feathers.

Kachinas, (Hopi men) masked to impersonate supernatural beings, present kachina dolls to Hopi children at kachina ceremonies. The dolls are not dolls in the usual sense. They serve an educational purpose: they help the children learn what the many different kachinas look like. The children also learn about the rituals.

## Materials

graph paper
instructions

## Object

The Hemis Kachina activity page provides practice in plotting given points and drawing line segments from point to point to reveal a hidden picture.

## Activity

The kachina of the picture graph represents the Hemis Kachina, also called the Jemez Kachina. It is frequently (and mistakenly) called the Niman Kachina because it is seen most often in the Niman or Home Dance.

The picture graph was drawn from a Hemis Kachina doll purchased in 1928 by a member of the writer's family at Keams Canyon, Arizona. The picture is drawn in proportion—head to body, tableta (headdress) to body, and legs to body. Further, the picture graph depicts the old style of the kachina doll: The arms were carved against the body. Today a kachina doll may be carved so that the kachina doll appears to be dancing.

The carved kachina doll is elaborately decorated, especially the tableta; however, for our graphing purposes the doll has enough detail

to be recognizable in a group of kachina dolls as a Hemis Kachina. After completing the picture graph, students may wish to draw feathers at the top and corners of the tableta.

### REFERENCE

Colton, Harold S. *Hopi Kachina Dolls with a Key to Their Identification.* Albuquerque: University of New Mexico Press, 1959.

Solution

# Hemis Kachina

**Plot these points on your graph paper:**

| | | | | | |
|---|---|---|---|---|---|
| (6,36) | (7,36) | (4,34) | (6,34) | (7,34) | (9,34) |
| (2,31) | (4,31) | (9,31) | (11,31) | (3,27) | (10,27) |
| (4,26) | (6.5,26) | (9,26) | (4,25) | (6,25) | (7,25) |
| (9,25) | (4,24.5) | (6,24.5) | (7,24.5) | (9,24.5) | (2,23) |
| (4,23) | (9,23) | (11,23) | (4,22.5) | (6.5,22.5) | (9,22.5) |
| (4,22) | (9,22) | (2,21) | (11,21) | (4,20) | (9,20) |
| (4,16) | (6,16) | (7,16) | (9,16) | (2,15) | (11,15) |
| (3,14) | (4,14) | (6,14) | (7,14) | (10,14) | (5,13) |
| (4,12) | (5,11) | (4,10) | (5,9) | (4,8) | (2,7) |
| (5,7) | (11,7) | (2,6) | (4,6) | (6,6) | (7,6) |
| (9,6) | (11,6) | (4,5) | (6,5) | (7,5) | (9,5) |
| (4,3) | (6,3) | (7,3) | (9,3) | (3,2) | (6,2) |
| (7,2) | (10,2) | | | | |

**Draw line segments from—**

| | | | | | | | |
|---|---|---|---|---|---|---|---|
| 1. (4,23) | to | (2,23) | 24. (2,15) | to | (3,14) | 47. (10,2) | to (9,3) |
| 2. (2,23) | to | (2,31) | 25. (3,14) | to | (6,14) | 48. (9,3) | to (9,6) |
| 3. (2,31) | to | (4,31) | 26. (6,14) | to | (6,16) | 49. (4,5) | to (6,5) |
| 4. (4,31) | to | (4,34) | 27. (6,16) | to | (4,16) | 50. (4,3) | to (6,3) |
| 5. (4,34) | to | (6,34) | 28. (4,16) | to | (4,20) | 51. (7,5) | to (9,5) |
| 6. (6,34) | to | (6,36) | 29. (9,22) | to | (11,21) | 52. (7,3) | to (9,3) |
| 7. (6,36) | to | (7,36) | 30. (11,21) | to | (11,15) | 53. (4,14) | to (5,13) |
| 8. (7,36) | to | (7,34) | 31. (11,15) | to | (10,14) | 54. (5,13) | to (4,12) |
| 9. (7,34) | to | (9,34) | 32. (10,14) | to | (7,14) | 55. (4,12) | to (5,11) |
| 10. (9,34) | to | (9,31) | 33. (7,14) | to | (7,16) | 56. (5,11) | to (4,10) |
| 11. (9,31) | to | (11,31) | 34. (7,16) | to | (9,16) | 57. (4,10) | to (5,9) |
| 12. (11,31) | to | (11,23) | 35. (9,16) | to | (9,20) | 58. (5,9) | to (4,8) |
| 13. (11,23) | to | (9,23) | 36. (3,14) | to | (2,7) | 59. (4,8) | to (5,7) |
| 14. (9,23) | to | (9,22) | 37. (2,7) | to | (2,6) | 60. (5,7) | to (4,6) |
| 15. (9,22) | to | (4,22) | 38. (2,6) | to | (11,6) | 61. (6.5,26) | to (6.5,22.5) |
| 16. (4,22) | to | (4,26) | 39. (11,6) | to | (11,7) | 62. (4,22.5) | to (9,22.5) |
| 17. (4,26) | to | (3,27) | 40. (11,7) | to | (10,14) | 63. (4,25) | to (6,25) |
| 18. (3,27) | to | (10,27) | 41. (4,6) | to | (4,3) | 64. (6,25) | to (6,24.5) |
| 19. (10,27) | to | (9,26) | 42. (4,3) | to | (3,2) | 65. (6,24.5) | to (4,24.5) |
| 20. (9,26) | to | (9,23) | 43. (3,2) | to | (6,2) | 66. (9,25) | to (7,25) |
| 21. (4,26) | to | (9,26) | 44. (6,2) | to | (6,6) | 67. (7,25) | to (7,24.5) |
| 22. (4,22) | to | (2,21) | 45. (7,6) | to | (7,2) | 68. (7,24.5) | to (9,24.5) |
| 23. (2,21) | to | (2,15) | 46. (7,2) | to | (10,2) | | |

# Totolospi

## Background

Totolospi, pronounced to-to-los-pi, is a Hopi game of chance that was played with two or three cane dice and a counting board inscribed on stone. Totolospi was played by adults as well as children.

Interestingly, totolospi has been described as a game of fox and geese and as resembling checkers, Monopoly, and Parcheesi. These diverse descriptions occur as the result of the variety in the stone totolospi counting boards acquired by different collectors.

Three varieties of totolospi counting boards are included: (1) an ellipse for two players, (2) two ellipses intersecting to form a cross for four players, and (3) a rectangular board for two or four players. The rectangular board is about the size of the inscribed red sandstone slab at the Southwest Museum in Highland Park, California. This board was drawn after measuring the board through the glass of the museum case. The cases are permanently sealed to protect the artifacts.

## Materials

counting board
2 or 3 can dice (Half-round dowels can be used for the cane dice. Cut each 9.5 cm in length. If desired, paint the round sides white and the flat sides red. If half-round dowels are not available, use Popsicle sticks.)
2 or 4 counters called "animals" (The Hopi used kernels of colored corn for the animals. Each player has a different color.)

## Number of players: 2 or 4

## Object

Totolospi is a game of chance. With older students, consider possible outcomes when two or three cane dice are used. Students can use tree diagrams as a useful tool for calculating probabilities. For younger students, the game provides practice in counting by ones and twos on the elliptical boards and by fives and tens on the rectangular board.

## How to play

*1. Totolospi for two:* Elliptical counting board

Each player places a counter (or animal) on the nearest circle. Alternate turns are taken. Movement is on the line segments of the board, not on the spaces between the segments. The moves are determined by holding three cane dice in one hand and dropping them on end. If three round (white sides) land up, the player advances 2 lines. If three flat (red sides) land up, the player advances 1 line. If a combination is thrown, the player loses his or her turn. The player reaching the opposite side first wins.

*2. Totolospi for four:* Cross-shaped counting board

Each player places an animal on the circle of the arm of the cross nearest to him or her. Players take turns. If three round sides land up, the player advances 2 lines. If three flat sides land up, the player advances 1 line. If a combination is thrown, the player loses his or her turn. The winner is the first to reach the opposite end of the arm he or she is playing on.

*3. Totolospi for two or four:* Rectangular counting board

For this version, movement is on the spaces rather than the lines. Players place their animals in the center circle, or "the house," the circle at the end of the strip within the rectangle. Movement is determined by dropping two cane dice. If both round sides land up, the player moves 10 spaces. If both flat sides land up, the player moves 5 spaces. A player continues to play as long as the cane dice show both sides alike. If the dice land with unlike sides, then the next player plays. *Two animals may not occupy the same space.* The second animal landing on a space sends the first one back to the house to begin again.

At the end of the inside path, the player must turn and continue around the board, called "the home road." The player may turn either right or left. At the last corner after going around the board, if the player's dice land flat sides up, he or she may go to the door of the house and later proceed inside to the starting point. If, however, the player's dice land round sides up, he or she has gone past the door and must go around the home road again. The winner is the first player to return to the starting place. If, after a second trip around the board, the player goes past the door of the house again, he or she loses.

(Continued)

**Totolospi** (Continued)

### After playing

With older students, consider the possible outcomes when two or three cane dice are used. To begin, if one cane is dropped, it may land with the round side up, (R) or the flat side up (F). The possible outcomes of dropping two cane dice are shown in figure 1.

|  | Second Cane | |
|---|---|---|
| | R | F |
| First Cane — F | FR | FF |
| First Cane — R | RR | RF |

Fig. 1

There are 2 x 2 possibilities. Since each outcome is equally likely to occur, the probability of a given event *P* can be calculated as follows:

$$P \text{ (an event)} = \frac{\text{number of outcomes players are interested in}}{\text{total number of possible outcomes}}$$

Therefore, the probability of the cane dice landing with two like sides up is 2 out of 4. This is written 2/4, or 1/2. The probability of 1/2, or .5, means that the chance of the dice landing with two sides alike is equal to the chance of the dice not landing with two sides alike.

However, in the rectangular totolospi board, the player needs two round sides up to move ten spaces. The probability of the cane dice landing with two round sides up is 1 out of 4, or 1/4. If the player should need five to go to the door of the house, the probability is the same. The probability of two flat sides landing up is 1 out of 4, or 1/4.

For the other two board games—the ellipse and the cross—three cane dice are used. The possible outcomes are listed in the tree diagram (see fig. 2). There are 2 x 2 x 2 possibilities.

Movement in these totolospi games is dependent on the dropping of three cane dice with three round sides landing up or three flat sides landing up. These events are mutually exclusive; that is, if all three land round sides up, this excludes three flat sides up or other outcomes.

The probability of three round sides landing up is 1/8. The probability of three flat sides landing up is 1/8. The probability of either three round or three flat sides landing up is determined as follows:

$$\begin{aligned} P \text{ (R or F)} &= P \text{ (R)} + P \text{ (F)} \\ &= 1/8 + 1/8 \\ &= 2/8 \end{aligned}$$

The probability of either three round or three flat sides landing up is 2/8, or 1/4. Ask students if they think that three cane dice landing with three round sides up or with three flat sides up is easy or difficult to obtain.

### REFERENCES

Bruce Bryan (Curator, Southwest Museum, Highland Park, Los Angeles), 1980: personal communication.

Watkins, Frances E. "Indians at Play: I—Hopi Parcheesi." *Masterkey* 18 (September 1944): 139–41.

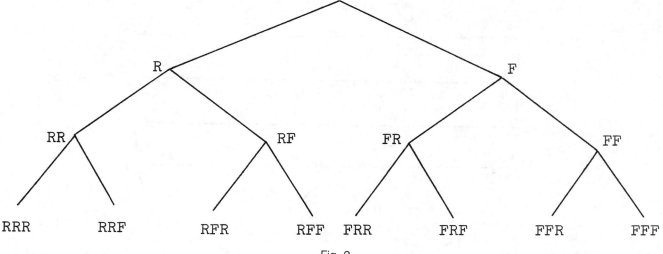

Fig. 2

73

# Totolospi

# Totolospi

# Totolospi